D1367763

Machine Pacing
and Occupational Stress

Machine Pacing
and
Occupational Stress

Proceedings of the International Conference, Purdue University,
March 1981

Edited by

Gavriel Salvendy

Purdue University, West Lafayette,
Indiana, USA

and

M. J. Smith

National Institute for Occupational
Safety and Health, Cincinnati,
Ohio, USA

Taylor & Francis Ltd
London
1981

First published 1981 by Taylor & Francis Ltd,
4 John Street, London WC1N 2ET

The material published in this book was written
under a contract of the government of the United
States of America and is published without copy-
right restrictions.

British Library Cataloguing in Publication Data

Machine pacing and occupational stress.
   1. Psychology, Industrial 2. Labor productivity
   3. Stress (Psychology)
   I. Salvendy, Gavriel II. Smith, M. J.
158.7   HF5548.8

ISBN 0-85066-225-7

Printed and bound in Great Britain by
Taylor & Francis (Printers) Ltd,
Rankine Road, Basingstoke, Hampshire RG24 0PR.
Phototypesetting by Georgia Origination, Liverpool.

Contents

Preface

This book consists of papers presented at the First International Conference on Machine Pacing and Occupational Stress held at Purdue University 17–19 March 1981. This conference was sponsored by:

Division of Biomedical and Behavioral Science,
National Institute for Occupational Safety and Health (NIOSH),
Centres for Disease Control,
Public Health Service
U.S. Department of Health and Human Services

and

Human Factors Program,
School of Industrial Engineering,
Purdue University.

The rationale for this international conference stems from the fact that there are, worldwide, over 50 million people working on machine-paced jobs. Such jobs enhance industrial productivity and, hence, increase our standard of living. Yet they are suspected of causing social unrest in the work environment and possibly ill-health. Because of the potential advantages and disadvantages of the paced-work environment, researchers are interested in studying such work with the notion of enhancing the human values of the paced-work environment without hindering productivity.

Stress is of interest to machine-paced work, since it encompasses a number of components (fatigue, arousal, attentive behaviour and mental load) which have been associated with workers' discontent of machine-paced work. Stress may be most appropriately defined operationally through measurable changes in physiological, psychological and performance variables. The fact that stress is suspected of being one of the major causes of mental ill-health, coronary heart disease and stomach disorders makes it a vital candidate for study.

The concepts for the conference were germinated during the May 1979 visit of Gavriel Salvendy to the Finnish Institute of Occupational Health in Helsinki, where Kari Lindstrom, Director of the Department of Psychology, and Ilka Kuorinka, Assistant Director of the Department of Physiology, expressed an interest in collaborative research with the USA on machine pacing and occupational stress. The fact that both NIOSH and Purdue University were engaged in a broad range of research studies in this area resulted in the initial organization of a joint USA–Finnish workshop. As this workshop developed, it became evident that a worldwide perspective should be included in such a venture. This led to the First International Conference on Machine Pacing and Occupational Stress in which 45 of the 47 papers appearing in this book were presented, representing the work of scientists from 12 nations.

It was recognised at the outset by the conference organizers that, for research findings to have a real impact on the workplace, the senior trade union and management personnel would have to participate. Following this notion, the Vice President of Labor Relations at the American Telephone and Telegraph Corporation and his counterpart at The Communications Workers of America addressed the conference. The conference organizers were especially pleased to see among the attendees practitioners representing many of the leading manufacturing, government and insurance organizations. The fact that 27 graduate students from seven major universities in the USA participated indicates both the current interest and the availability of future research manpower in this area.

Talking about the importance of applying our research findings in the workplace, Barry L. Johnson, Director of the Division of Biomedical and Behavioral Science, welcomed the participants on behalf of NIOSH. An extract from this welcoming address is presented below.

Within the context of our activities in stress alleviation, we feel the necessity to investigate the efficacy of self-administered procedures that may reduce stress at the workplace. Such procedures may include muscle relaxation, biofeedback procedures and so forth. [We] feel compelled to mention our research in stress reduction, owing to the rather wide distribution of a recent article that questioned the involvement of NIOSH in transcendental meditation. While we are not into transcendental meditation, *per se*, we do view, as I mentioned, the need to perform research on means for stress reduction. To fail to perform such research would abrogate our responsibilities under the OSH Act. More importantly, we will have failed those whom we are to serve.

Felix Haas, Executive Vice President of Purdue University, welcomed the participants on behalf of the University.

The organization of this conference and the publication of this book was made possible through the help of many people. It was a most rewarding experience to receive all the chapter manuscripts on time and within the very tight, allocated length of ten manuscript pages. For this, our sincere thanks goes to the chapter authors.

We are grateful that much of the work for the organization of the conference and the compilation of this book was undertaken by Joyce Hinds, Head Secretary, who was diligently assisted by Becky Austin, Nancy Cassel, Wanitta Stutzman and Linda Wann.

West Lafayette, Indiana, May 1981   *Cincinnati, Ohio, May 1981*
G. Salvendy   *M. J. Smith*

Section 1. Perspectives on work pacing and stress

Comments from the sidelines

E. J. McCormick

Department of Psychological Sciences,
Purdue University, USA

Many manifestations or indicators of work-related stress have been reported, including physiological and physical indicators, psychological and attitudinal indicators, effects on job performance, and accidents and injuries. Further, there is evidence that work-related stress is very pervasive, extending over the entire spectrum of the relationship of people to work, starting with the employment process, continuing through training, and occurring in virtually all types of job; in addition, it is sometimes associated with the interpersonal relationships of workers with their supervisors or with other workers, with certain characteristics of organizations, and with environmental variables. The objectives of stress-related research probably should be focused on its reduction to acceptable levels rather than towards its elimination.

On the first day of a class in Industrial Psychology that I taught I announced that this was a class in Industrial Psychology. A surprised student spoke up and said, 'You said this is a class in Industrial Psychology? I am supposed to be in a class in Horticulture!' In one sense I feel as out of place here as the horticulture student must have felt, especially being up here talking to you, since I am pretty much of a non-expert in the field of stress, surrounded by an assemblage of some of the world's experts on stress.

Although I can make no pretense of being an expert on stress, in my professional activities over the years I have encountered, or have been involved in, scads of varied circumstances in which some aspect of stress has surfaced. These experiences have impressed upon me the pervasiveness of stress in its various forms and manifestations across the whole spectrum of the relationships of people with their work activities. I might add that my own professional background generally has covered the three areas of Industrial Psychology, Human Factors, and Job Analysis. In all of these areas references to stress have surfaced in a wide variety of situations. The indicators or manifestations of possible stress that keep bobbing up typically can be viewed as the basis for the measurement of strain, that is, the consequences of stress. These indicators tend to fall into four groups as follows.

(1) Physiological and physical indicators such as: heart rate; heart-rate recovery rate; sinus arrythmia; oxygen consumption; oxygen debt; galvanic skin response; ECG; electromyograms; muscle tension and muscle disorders; blood pressure; chemical changes in the blood; blood sugar; perspiration; and respiration rate.

(2) Psychological and attitudinal indicators such as: boredom; job dissatisfaction; aggressiveness; apathy; hypertension; nervousness; psychosomatic disorders; and sensations of fatigue.

(3) Job-related behaviours such as: the quality and quantity of work performance on primary and secondary tasks; reaction time; variability in cycle time (on jobs that have a cycle); inadequate timing of responses; lack of attention or alertness; tardiness; absenteeism; and job turnover.

(4) Accidents and injuries.

I should add that these indicators or manifestations are by no means the exclusive result of stress as such. Certain other job-related and individual-related variables also contribute to some of these indicators.

Now let me refer to a limited sample of job-related situations in which some indicators of stress have been reported. We can start with the time people apply for jobs. In taking employment tests, for example, some people tend to panic and fail to perform as well as they might. And in employment interviews some people clam up, or become nervous and fail to convey an appropriate impression of their potential. It might be added that some organizations use what are called stress interviews in which job candidates are intentially subjected to interpersonal stress by the interviewer in order to gain some insight into the candidates' abilities to cope with such stress. For those who are employed, the tensions some people experience during training tend to impede learning or cause people to fall by the wayside.

The stress associated with various types of actual jobs has been rather widely documented. Stress associated with decision-making and other mental activities, for example, has been reported for various executive and management types of job, computer operations, aircraft control operation, and certain types of office work. In such instances the precipitating job variables include mental workload, attention demands, and various types of job 'pressures', such as time pressures. In the case of essentially physical jobs the stressful variables can arise from physical workload, attention demands, maintenance of a workpace, the repetitive nature of jobs, or other features.

In a more general frame of reference there is growing concern regarding the trend towards increased computer control of both office types of operation ('word-processing' work is now the common expression) and of production-type jobs. Two points of view regarding this trend seem to have emerged. On the one hand, it is argued that increased computer control can make jobs more congenial by eliminating the repetitious tasks that can be handled by computers, so people can serve as 'exception handlers', solving the problems the computer cannot, thus increasing the responsibility of workers. On the other hand, it is argued that this trend will result in jobs becoming more and more routine and simplified, with workers losing 'control' of their jobs, thus becoming more the puppets of computer-controlled systems. If increased automation has this effect, it is probable that increased work-related stress will be inevitable. I will leave it to others to predict which way the wind will blow, but I do foresee this as an issue of some consequence.

Aside from job-related variables serving as sources of stress, the inter-personal relationships of people also can serve to generate stress. This is particularly the case with supervisor–subordinate relationships. This actually can be a two-way stress—either a street with chuck-holes that generate stress on the part of either or both individuals, or one with a smooth pavement that makes for a satisfying personal interaction. In addition, the interpersonal relationships of fellow workers can operate in both directions.

Still another aspect of the working situation that can serve as a source of stress is the nature of the organization and of its practices, policies, and 'climate'. The prevailing climate or 'atmosphere' of an organization—its 'ambience'—in large part is generated by management, and can have a chain-reaction effect within the organization, either towards making the organization a satisfying place to work, or by generating conflict, friction, anxiety, and dissatisfaction (which in turn can serve as a source of stress for some people). In addition, the practices of the organization regarding employee participation, the latitude of decision-making, and latitude regarding job activities also can contribute to, or minimize, stress on the part of employees.

Further, various features of the physical work environment can serve as stressors, in particular noise, temperature and humidity, poor illumination, and in some instances vibration.

We could cite other types of job-related situation that have been found to contribute to stress, but these examples serve to bring me to the central point that I would like to make, namely that stress is a very pervasive, wide-spread, phenomenon that can occur in virtually all aspects of the relationships of people to their work. Sitting on the sidelines of your collective domains of stress, it seems to me (as a non-expert) that stress is as broad as it is wide. Not only can it raise its head in many types of job-related situation, but it can be manifested in many different ways, as implied by the variety of indicators or criteria of strain that I referred to above. There may well be some syndrome of human physiological responses that are common to stress, but its tentacles can influence a variety of indicators of strain, including physiological reactions, psychological and attitudinal reactions, effects on job performance and job attendance, and accidents and injuries.

Although the major points that I have had in mind in my comments have been to call attention to the pervasiveness of stress in the many facets of the world of work and to its many manifestations, I would like to add a few additional reflections about stress and its effects.

In the first place I am inclined towards the view that there is no completely 'non-stressful' work situation, so the notion of eliminating sources of stress is probably a will-of-the-wisp. In fact, there are various types of data and hypotheses to suggest that, for at least certain types of work, there might be some level of some form of stress that might be 'optimum' in terms of work performance. The inverted-'U' hypothesis, which has been around for some time, has been applied to such stress-related factors as muscle tension, alertness, and sensory stimulation. According to the hypothesis, both low and high levels of the stress-related factor are associated with poor performance, intermediate levels being associated with 'optimum' performance. Thus, the objectives of efforts to 'do something' about stress should probably be aimed at achieving an 'optimum' level (if there be such an optimum for a given situation) or at least to achieve an acceptable level.

Following on from this point, it is relevant to refer to the possibility of modifying work situations to minimize, or to 'control', the level of some sources of stress. In this regard it can be said that it is possible to take such action in the case of at least certain possible sources of stress, such as by the application of human factors or ergonomic principles and data to job design, and by modifying organizational and interpersonal relationships. However, the present realities of the world of work are such that there would be limits to such efforts—that is, the nature of some types of job is such that certain types of stress cannot be eliminated nor even reduced.

This leads us to another point that I would like to emphasize, namely that we need to recognize the fact of wide individual differences in susceptibility to stress. The determination of the personal characteristics that are related to susceptibility to stress would make it possible to identify those individuals who most likely would be able to tolerate the specific stresses of particular types of work. An emphasis on personnel selection along these lines would be particularly relevant in the case of work situations in which it is not feasible to bring stress levels down to a level that would be tolerable for virtually all possible job candidates.

In summary, I would like to reinforce the point that stress seems to me to be a very pervasive factor in work situations, and that its manifestations are indeed varied and complex. Granting that it is unrealistic to consider the elimination of work stress, and probably not even desirable, it seems to me that the objectives of research relating to stress should be aimed at efforts to bring it within reasonable bounds. This objective would seem to be especially relevant in the case of those aspects of stress that have the most serious implications in terms of the welfare of people and of the effectiveness of their work activities.

Classification and characteristics of paced work†

By G. SALVENDY

Human Factors Program
School of Industrial Engineering
Purdue University
West Lafayette, Indiana 47907
USA

The advantages and disadvantages of machine-paced (M/P) work are reviewed and a classification of paced work is presented. Based on experimental evidence, the following results are discussed: (1) performance feedback reduces stress; (2) short rest periods between work cycles reduce stress for tasks requiring external attention, but not for tasks requiring internal attention; (3) no statistically significant, or clinically meaningful, differences exist in psychophysiological indicators of stress between M/P and self-paced (S/P) work; (4) no statistically significant differences exist in the quality of work performance between M/P and S/P work; and (5) the psychological attributes of those who prefer M/P work and those who prefer S/P work are presented.

1. Introduction

Various subjective estimates suggest that, worldwide, over 50 million people are working on M/P tasks. Hence, improving the working conditions on M/P tasks has spurred the interest of many researchers. In effect, over 100 scientific papers have been published on this subject (Salvendy and Burke 1981). However, there is a high degree of diversity in the results obtained from these experiments. It is quite common to find a situation where results obtained from one study are completely contradicted in another study. This makes it very difficult to draw inferences which could be used to improve the working conditions on M/P tasks.

This suggests the need for the classification of paced work and the classification of M/P research. An early attempt in this area was made by Murrell (1965). In this chapter, the advantages and disadvantages of M/P work are reviewed, and a classification of M/P work is proposed. Following this, some of the studies performed by the author and his associates on M/P and S/P tasks are discussed.

† The preparation of this chapter was facilitated by the financial support from the National Science Foundation (Grant APR7718695) and National Institute of Occupational Safety and Health (Contracts Nos. 210-80-0034 and 210-80-0002).

2. Who needs machine-paced work?

If there had not been a distinct economic advantage in utilizing M/P work, there may not have been over 50 million people working in this area. It also would be true to say that 'if there had not been some disadvantages for the human working on M/P tasks, there may not have been over 85 publications in this area.

The economic advantages and disadvantages of using M/P tasks are reviewed in table 1. The frequently referred to psychological disadvantages of M/P work include the following: (1) M/P work does not provide psychological growth for the workers; and (2) M/P work causes boredom and job dissatisfaction.

Table 1. Economic advantages and disadvantages of machine-paced work.

	Advantages
(1)	Reduces overhead cost through: economic use of high technology; reduction of stock in progress; reduction in factory floor space; reduction in supervision cost.
(2)	Reduces direct cost through: decreased training time; lower hourly wages; high production return per unit of wages.
(3)	Contributes to national productivity through: provision of employment for less capable workers; reduction in the production costs of goods and services.
	Disadvantages
(1)	Does not have provision for the utilization of each worker's maximal work capacity.
(2)	Economically viable only for high-volume production.

We should only maintain M/P tasks in the work environment if we can simultaneously maintain the economic benefits and alleviate the human disadvantages of working on an M/P task. In order to accomplish this, we must embark on systematic and structured research to alleviate the negative aspects of M/P tasks. For this, a classification of M/P work is needed.

3. Classification of paced work

M/P work can either be classified by the demand it places on human behaviour and performance (table 2) or by the research methodologies utilized to study M/P work (table 3). Both classifications are needed in order to effectively integrate and implement the research findings pertaining to M/P work. In addition to the classifications (tables 2 and 3), three items, pertaining to job content, must be considered, namely: (1) The *perceptual and mental load* associated with task performance—the higher the load the higher the psychophysiological stress associated with task performance; (2) the *decision*

Table 2. Classification of paced work.

Human-paced work

(1) *Truly unpaced*—No internal or external pacing is imposed. The task is performed at a preferred and chosen pace by the operator.

(2) *Socially-paced*—Although no pacing is imposed by management or by machinery, there is a peer or group pressure to perform to a set pace. Examples are group performance and lectures.

(3) *Self-paced*—Although no machine pacing is present, the work is paced by management objectives. An example is day rate work, where an operator must produce by the end of the workday a specified number of items. In this example, provided that by the end of the day the operators produce the required amount, they may spend differing amounts of time on each cycle of the job.

(4) *Incentive-paced*—An incentive-paced task consists of two additive parts, namely the 'self-paced' component and the operator's financial motivation to produce above the self-paced work. The more the operator produces above this 'self-paced' level, the higher will be the operator's income. Hence, the intensity and the severity of the pacing is dictated by how much the operator wants to earn.

Machine-paced work

(1) *Length of work cycle*—When the cycle length in machine-paced work is extremely long, it approaches the state of the 'self-paced' condition. The shorter the cycle time, the less the operator's performance variability can be tolerated.

(2) *Buffer stocks*—Buffer stock is 'an arrangement which makes more than one component of feeding position available to an operative at the same time' (Murrell 1965). Machine-paced work can be operated with or without buffer stocks. When machine-paced work is designed with a buffer between stations, the effects of stringency associated with machine-paced work may be reduced; the larger the buffer stocks, the smaller the stringency associated with machine-paced work. Extremely large buffer stocks between work stations may reduce the effects of machine-paced work to those of self-paced work.

(3) *Rate of machine-paced work*—When a fair day's work is defined as 100%, the rate of M/P work is frequently performed at rates ranging from 100–125. The impact of M/P work on the operator may be different, depending at which rate the task is performed.

(4) *Continuous versus discrete pacing*—Both pacing modes are widely utilized in industry. For example, in conveyor operations the conveyor can either move continuously, in which case the operator performs the task in a dynamic visual work environment, or the conveyor can be indexed in a discrete mode. In the latter case, the conveyor is in a stationary mode during a fixed job cycle period when the operator is typically working on the job. At the end of each work cycle, the conveyor indexes to the next workstation. During this indexing period (which usually takes 2–8 seconds), the operator can either be doing preparatory work for the next cycle of operation or be idle.

Table 3. Classification of machine-paced research.

---

(1) *Laboratory studies*
    Typically performed for very short work period (i.e., less than for one full work day); on non-realistic tasks; on operators who are insufficiently experienced in task performance. Although laboratory studies are typically characterized by highly controlled experiments, the above-listed weaknesses of M/P laboratory studies make the transfer of research findings to real-world work situations suspect.

(2) *Epidemiological studies*
    Frequently in these studies, jobs are confounded with workers since it is extremely difficult to have in epidemiological studies a statistically balanced design. For this reason, epidemiological studies must be interpreted with extreme caution.

(3) *Confounded industrial studies*
    In these studies M/P is compared with S/P work; however, neither the operators nor the job content in the two pacing conditions are the same. Hence, in these studies pacing mode is confounded with job content and operators. This makes it very difficult to make comparative statements regarding M/P and S/P work.

(4) *Controlled industrial studies*
    In these studies the job content is the same for both M/P and S/P work and operators perform, in a statistically balanced experimental design, both pacing modes. This mode of studies enables the best transfer of knowledge to real-world work situations. This methodology is by far the most powerful of the four considered research methodologies.

---

latitude associated with task performance—the higher the decision latitude the lower the stress level associated with task performance; in effect, it is the interaction between job demand and job decision latitude that is the determinant of the level of stress in work situations (Karasek 1979); and (3) *performance feedback*, which reduces stress. The more precise the performance feedback, the lower the stress associated with task performance (Knight and Salvendy 1981).

4. Characteristics of paced work: some research findings

From the research findings of the author and his associates relating to the comparative merits of M/P and S/P work, the following emerges.

(1) It becomes evident when utilizing young subjects on a pump ergometer, a bicycle ergometer and a Harvard step test that the human body's efficiency is higher in S/P than in M/P work (Salvendy 1973); however, for older subjects, using an arm ergometer, the highest efficiency occurred in M/P work, whereas on the same task for younger subjects, the highest efficiency occurred in S/P work (Salvendy and Pilitsis 1971).

(2) Perceptual load associated with task performance plays a significant

role in evaluating M/P and S/P tasks. When the perceptual load of the task is low, there are no significant differences in stress levels between M/P and S/P tasks; however, for tasks with high perceptual load the stress is significantly lower during M/P than S/P task performance and the error rates are higher (Salvendy and Humphreys 1979). These results are attributed to two facts, namely: (1) high stress during task performance is associated with high need of achievement motivation, high production performance and low error rate (Salvendy and Stewart 1975); and (2) during the S/P task the operator had to keep track of the quantity of work output, whereas in M/P tasks the work output was controlled by the machine. It was hypothesized that this additional task of keeping track of work output imposes additional mental load and increases the stress associated with high mental load task performance. In a study by Knight and Salvendy (1981), subjects performed a task with high perceptual load both in the M/P and S/P modes. In the S/P mode the subjects performed the task with a variety of different performance feedbacks. Table 4 shows that the stress associated with task performance is the function of performance feedback. The more precise the performance feedback, the lower the stress associated with task performance.

Table 4. Effects of performance feedback on the reduction of stress in self-paced work.

Work condition	Stress index*
Self-paced	
No feedback	100
Cycle feedback	90
Time feedback	86
Combined time and cycle feedback	57
Machine-paced	62

*A difference in the stress index of six or more units is statistically significant at 5% level.

(3) In man–computer interactive work, the attentional work environment has a much greater impact on the stress associated with task performance than the stresses associated with M/P and S/P task performance. This is illustrated in table 5. In this experiment (Sharit and Salvendy 1981), subjects were asked to perform the same task with both M/P and S/P, in both financial and non-financial work environments. A waiting, or anticipation, period of 4–6 seconds was introduced between each work cycle. The physiological measures presented in table 5 reflect on the deceleration and acceleration of the heart-beats during the waiting period. The external attention task required visual input from a VDT terminal, whereas the internal attentional task required arithmetic calculations.

Table 5. Acceleration and deceleration in heartbeats as a function of job content, job design and the nature of incentives.

| | External attention | | Internal attention | |
	M/P	S/P	M/P	S/P
No incentive	66	63	71	69
Financial incentive	68	62	75	72

Mean starting heartbeat 72 beats per minute.

(4) In an industrial study (Salvendy 1981), 36 female operators performed, in a statistically balanced design, the same task at the same work output, both in the M/P and S/P modes, in each mode for a two-week continuous period. During this period, psychophysiological variables were unobtrusively continuously monitored (Knight *et al.* 1979, 1980 a and b) and a battery of psychological tests were administered. The results of this experiment support the following hypothesis.

(*a*) None of the psychophysiological measures indicated either statistically significant or clinically meaningful differences between M/P and S/P work. In effect, table 6 illustrates that five out of the six variables measured had less than 1% difference in their mean values between M/P and S/P work and one variable had a 3% difference.

(*b*) During the performance of 327 120 soldering operations, there were no statistically significant differences in the quality of work between M/P and S/P work.

(*c*) The psychological characteristics of those operators who prefer to work at M/P and those who prefer to work at S/P are shown in table 7. Two-thirds of the studied working population preferred S/P to M/P work; but about one-quarter prefered M/P to S/P work. However, as in a previous study (Salvendy 1978), about 10% of those studied did not like any work.

Table 6. Mean values of the psychophysiological data derived during the continuous monitoring of 36 industrial operators performing the same task in both M/P and S/P mode.

Psychophysiological variables	S/P	M/P	Percentage difference of M/P from S/P
Heartbeats (per minute)	78·93	79·53	+0·8
Sinus arrhythmia	100·57	101·00	+0·4
Rate of breathing per minute	12·01	11·63	−3·2
Mean blood pressure (BP) (at ankle)	117·93	117·33	−0·5
Systolic (BP) (at ankle)	146·60	145·60	−0·7
Dyastolic (BP) (at ankle)	100·75	101·66	+0·9

Table 7. Psychological profiles of operators who prefer S/P and those who prefer M/P work.

M/P work	S/P work
Less intelligent	More intelligent
Humble	Assertive
Practical	Imaginative
Forthright	Shrewd
Group-dependent	Self-sufficient

5. Is a comparative study between M/P and S/P work desirable?

The answer to this question is a definite NO. There are a number of reasons for this, among which the following may be mentioned.

(1) In real-world work situations, M/P and S/P work are most frequently analogous to simplified and enlarged jobs. Hence, in real-world situations these two pacing conditions are confounded with job content. In order to alleviate this confounding effect, researchers have attempted to match the job contents for the two pacing conditions. By so doing, artificial work situations emerged.

(2) In real-world situations, operators are engaged in either M/P or S/P work. Yet when a statistically balanced experimental design is performed, the operators have to perform at both pacing conditions. However, operators' gained experience predominantly on only one pacing mode; hence, prior experiences are confounded with pacing mode.

Rather than expanding research efforts which are aimed at comparing M/P with S/P work *per se*, a more fruitful avenue of research may be aimed at a better understanding of human behaviour (of which human stress is a component). Based on this understanding, a development of work structures within organizations may be achieved which improves human behaviour and performance. By so doing, through the modelling of the operator in work situations, a theoretically sound basis for work design could be established, rather than patching and correcting the many non-scientifically based work designs in order to attempt effectively to accommodate human behaviour and performance.

References

KARASEK, R. A., 1979, Job demands, job decision latitude, and mental strain: implications for job redesign. *Administrative Science Quarterly*, **24**, 285–308.
KNIGHT, J. L., and SALVENDY, G., 1981, Feedback effects in externally and internally paced tasks. *Ergonomics*, in the press.

KNIGHT, J. L., SALVENDY, G., and GEDDES, L. A., 1979, A minicomputer system for long-term automatic blood pressure monitoring. *Annals of Biomedical Engineering*, **7**, 369–374.

KNIGHT, J. L., GEDDES, L. A., and SALVENDY, G., 1980a, Continuous unobtrusive, performance and physiological monitoring of industrial workers. *Ergonomics*, **23**, 500–506.

KNIGHT, J. L., SALVENDY, G., GEDDES, L. A., JANS, K., and SMITT, E., 1980b, Monitoring the respiratory and heart rate of assembly-line factory workers. *Medical and Biological Engineering and Computing*, **18**, 797–798.

MURRELL, K. F. H., 1965, A classification of pacing. *International Journal of Production Research*, **4**, 69–74.

SALVENDY, G., 1973, Physiological and psychological aspects of paced and unpaced performance. *Acta Physiologica*, **42**, 267–275.

SALVENDY, G., 1978, An industrial engineering dilemma: simplified versus enlarged jobs. In *Production and Industrial Systems: Future Development and the Role of Industrial and Production Engineering*, edited by R. Maramatsu and N. A. Dudley (London: Taylor and Francis).

SALVENDY, G., 1981, Physiological and psychological basis for the design of machine-paced work. Unpublished report, Human Factors Program, School of Industrial Engineering, Purdue University, West Lafayette, Indiana.

SALVENDY, G., and BURKE, W., 1981, *Machine-paced research: review and reappraisal.* Unpublished report, Human Factors Program, School of Industrial Engineering, Purdue University, West Lafayette, Indiana.

SALVENDY, G., and HUMPHREYS, A. P., 1979, Effects of personality, perceptual difficulty and pacing of a task on productivity, job satisfaction, and physiological stress. *Perceptual and Motor Skills*, **49**, 219–222.

SALVENDY, G., and PILITSIS, J., 1971, Psychophysiological aspects of paced and unpaced performance as influenced by age. *Ergonomics*, **14**, 703–711.

SALVENDY, G., and SHARIT, J., 1982, Occupational stress. In *Handbook of Industrial Engineering*, edited by G. Salvendy (New York: John Wiley) in the press.

SALVENDY, G., and STEWART, K., 1975, The prediction of operator performance on the basis of performance tests and biological measures. *AIIE Transactions*, **7**:4, 379–387.

SHARIT, J., SALVENDY, G., and DEISENROTH, M. P., 1981, Psychophysiological effects of machine-paced work in relation to motivational, external and internal attentional requirements of a task. *Ergonomics*, in the press.

Occupational stress: an overview of psychosocial factors

By M. J. SMITH

Applied Psychology and Ergonomics Branch,
Division of Biomedical and Behavioral Science,
National Institute for Occupational Safety and Health,
Cincinnati, Ohio 45226, USA

This paper presents a noncritical overview of the literature dealing with psychosocial aspects of occupational stress. A variety of workplace stressors that have been related to negative health consequences are discussed. In addition, three studies that have looked at the relative risk of stress problems across occupations are reviewed. The paper concludes with a discussion of future needs in occupational stress research.

1. Introduction

Stress is something that we all experience and that we can all relate to, even though we have a hard time defining what we mean. Stress has been variously defined as the cause of health problems and alternatively as the health problems themselves. For the purpose of this paper a model of psychosocial stress and disease proposed by Levi (1972) will be used. The model defines stress as the psychological or physiological reactions to environmental or psychosocial stimuli (stressors). Using this model, a variety of psychosocial stimuli can be examined which have the potential to be stress-producing.

This paper will provide an overview of research findings concerning occupational stress in terms of job features and specific occupations that have been shown to be stressful. There have been recent reviews of occupational-stress research that provide detailed examinations of these issues and these reviews should be consulted if more detail is desired (Cooper and Marshall 1976, Cooper and Payne 1978, Smith et al. 1980). Specifically, this paper will cover two topics, stressful job factors and high-stress occupations. A brief review of various job factors (physical working conditions, organizational features, workload, work hours, work role and work tasks) that have been identified as workplace stressors will be presented first. A more detailed review of studies that have comparatively evaluated various occupations will follow.

2. Job factors

Physical working conditions have a role to play in job stress which is typically not a primary, but more of an additive influence. Their main psycho-social impact has been shown to be on lowering worker tolerance to other stressors and on worker motivation (Caplan *et al.* 1975; Smith *et al.* 1978). Factory work environments that have excessive noise, poor ventilation, inadequate lighting, and hot or cold shop floors can create both physical and attitudinal problems (Caplan *et al.* 1975, Smith *et al.* 1978). Such problems are not limited to factories; they also occur in modern office buildings. In many offices, workers are packed together with noisy machinery and poor lighting, all of which have a similar effect to those found in factories (Smith *et al.* 1981). Essentially, physical working conditions make the individual worker more susceptible to influence by other sources of stress.

Two organizational factors have been shown to be of special significance for increased job stress and decreased worker health. These are: (1) job involve-ment or participation; and (2) organizational support, as reflected by supervisory style, support from managers and career development. Lack of participation in work activities has been demonstrated by Margolis *et al.* (1974), French and Caplan (1970) and Smith *et al.* (1981) to result in an increase in negative psychological mood. In terms of organizational support, Quinn and Shepard (1974) and Smith *et al.* (1981) have shown that close supervision and a supervisory style characterized by constant negative performance feed-back were related to high levels of stress and poorer worker health. Excessive impersonal monitoring of employee performance has been introduced by modern technology and can produce a management style that is stress-producing (Smith *et al.* 1981). Smith *et al.* 1981 have demonstrated that worker feelings of lack of involvement are related to stress and potentially to health complaints.

Career development is another major organizational stressor that has been studied. Concern over chances for promotion has been shown to be a significant stressor for office workers (Smith *et al.* 1981), while being passed over for promotion has been related to increases in both job stress and ill-health (Arthur and Gunderson 1965).

Security is the other side of career development. The threat of job loss is a very potent stressor. It has been tied to serious health disorders such as ulcers, colitis, severe emotional stress and patchy baldness (Cobb and Kasl 1977) as well as to increased muscular and emotional complaints (Smith *et al.* 1981).

Mental workload factors, such as quantitative underload/overload and workpace, and stress have an interesting relationship in that a balance is required to avoid negative health consequences. Underload is just as bad for your health as overload (Frankenhaeuser and Gardell 1976) in that it can affect both psychological wellbeing and hormone activity. Sedentary jobs create habit patterns that carry over into other aspects of life, thus influencing health risk off the job.

Quantitative overload has been shown to be a significant stressor for various occupations including scientists (French and Caplan 1970), machine operators

(Caplan *et al.* 1975) and data-entry clerks (Smith *et al.* 1981). The impact of work overload varies from psychological disturbances (Margolis *et al.* 1974, Smith *et al.* 1981) to increased disease risk (French and Caplan 1970, Quinn and Shepard 1974, Margolis *et al.* 1981, Smith *et al.* 1981).

Workpace is a very important workload factor. The speed or rate of work has been implicated as a significant issue in factory-worker ill-health (Murphy and Hurrell 1980, Hurrell 1981). Recent technology, such as computers, that can operate at high speeds on a continuous basis has increased the pacing impact on office workers. Recent research suggests that pacing produced by computerization may have an even greater effect than factory pacing (Cakir *et al.* 1978, Smith *et al.* 1981).

More and more operations are functioning in other than the typical daytime hours of work. This is particularly true of service operations and new technology such as computer installations. Shift work, particularly night and rotating shift regimens, has been shown to have a significant impact on health and safety, affecting such diverse areas as industrial injury incidence, worker sleeping and eating patterns, and family and social-life satisfaction (Rutenfranz *et al.* 1977, Harrington 1978, Smith *et al.* 1979, Smith *et al.* 1981).

Overtime is another factor that has been implicated in stress and ill-health. While there are only a few studies concerned with overtime health effects, these indicate a relationship between working 50 or more hours per week and coronary heart disease risk as well as psychological disfunction (Breslow and Buell 1960, Russek and Zohman 1958, Margolis *et al.* 1974). Overtime also has the potential to impact health in that it takes time away from family and friends and thereby reduces positive social interaction that can buffer stress.

Time pressure, such as having to meet deadlines, is another stressor that may interact with both work hours and workpace. Studies have shown increases in stress level as difficult deadlines draw near (Friedman *et al.* 1958, Caplan *et al.* 1975).

Work role includes a number of job factors such as responsibility for others, job conflict, role ambiguity, accountability, authority, discretionary control, participation and job status. A number of studies have demonstrated that role ambiguity, job conflict and responsibility for persons are related to job stress and psychological problems (Margolis *et al.* 1974, Caplan *et al.* 1975, Smith *et al.* 1981). The lack of discretionary control over work activities has also been shown to be related to increased risk of coronary heart disease (Karasek *et al.* 1981). Lack of participation in work decisions is yet another source of increased worker stress (French 1963, Caplan *et al.* 1975).

Work-task factors that have been researched include task variety, task clarity (confusion), challenge, complexity, utilization of skills and abilities, and activity level. All of these factors have been related to increased stress and negative psychological states such as boredom, confusion and frustration and have also been related to increased risk of health disorders (Frankenhaeuser and Gardell 1976, Caplan *et al.* 1975, Margolis *et al.* 1974, Smith *et al.* 1981).

Breaking work down into simple units, to reduce memory work and increase the pace of processing, produces a loss of skill and has brought about low-satisfaction jobs with high-stress levels and poor worker health (Caplan *et al.* 1975, Margolis *et al.* 1974, Smith *et al.* 1981).

3. High-stress occupations

This part of the review examines only research conducted in the USA that has evaluated occupations that are more likely to be stressful. There have been three major efforts which have comparatively evaluated a variety of jobs in terms of stress load. Quinn and Shepard (1974) surveyed approximately 1500 USA families, whose employed members represented a national probability sample utilizing a structured interview procedure. Many aspects of the individual's working life were evaluated for 12 different occupational groups studied, covering a range of white- and blue-collar jobs.

Responses to the job-stress and strain questions varied considerably across occupations. Professional and technical workers and managers scored highest in perceived job satisfaction but also showed high levels of depressed mood. Machine operators scored lowest for job satisfaction and low in perceived health, but scored the best of any groups on the mental-health measures. Labourers showed the second-worst job-satisfaction level, but perceived their health as good and scored well on mental-health measures. Overall, white-collar workers showed much greater job satisfaction than blue-collar workers although they showed slightly higher depressed mood and slightly poorer perceived health.

Caplan *et al.* (1975) conducted a major study, dealing with defining high-risk occupations and evaluating the impact of particular stressors. This study examined 23 occupational groups from over 2000 workers that were surveyed via questionnaire. As with the Quinn and Shepard (1974) study, the results indicated strong occupational differences in stress/strain levels. Specific stressors such as low utilization of abilities, lack of participation, work complexity, responsibility for persons, and role ambiguity were high for assembly-line workers, fork-lift drivers, and machine operators; but low for professors, family physicians, and other professionals. Machine-paced assembly-line workers scored high on boredom and dissatisfaction with work-load. The most satisfied occupational groups were family physicians, professors, and white-collar supervisors. Overall, assemblers and relief workers on machine-paced assembly lines had the highest levels of stress/strain.

A third study was a records evaluation of over 22 000 cases of stress-related disorders in 130 occupations in the state of Tennessee (Smith *et al.* 1980, Colligan *et al.* 1977). The results indicated that 40 of the 130 occupations had a stress-related disease incidence significantly higher than expected. Of these 40 occupations, 12 had very high incidence rates. These were labourers (general and construction), secretaries, inspectors (assembly-line), clinical-laboratory technicians, office managers, managers/administrators, foremen, waitresses/waiters, factory-machine operators, mine-machine operators, farm owners, and house painters. There were 77 occupations that showed expected stress-disease incidence and 13 occupations that showed significantly lower-than-expected stress-disease incidence. The 13 lower-than-expected incidence occupations included seamstresses, checkers/examiners, stockhandlers, freight handlers, craftsmen, maids, farm labourers, heavy-equipment operators, child-care workers, packers/wrappers, college professors, personnel/labour-relations workers, and auctioneers/hucksters. Some of these occupations were very similar in job requirements to the high-incidence occupations.

The three studies demonstrate the complex nature of the relationships between job features and health risk. While they have helped to define and verify critical workplace factors related to stress such as machine pacing, lack of control, poor supervisory relations, and lack of social support, they have not increased our understanding as to why certain job features are stress-producing.

4. Conclusion

The purpose of this review has been to provide a noncritical overview of job features and occupations that have been shown to be stressful in various studies of job stress. No attempt has been made to evaluate the adequacy of definitions, research design or the interpretation of reported results. Rather, the intent was to demonstrate that a wide range of job features have been examined and identified as being problematic. The number and variety of job features that appear to be stress-producing is large, and it is obvious that there is a need for further research. Machine-paced work is a primary example of the difficult job that must be undertaken when an occupational stressor is investigated. At the 1981 Purdue Conference on Machine Pacing and Stress, a wide variety of methodologies were presented, from field surveys to laboratory investigations, dealing with defining the health impact of machine pacing. When the findings from these various methodologies are integrated, there will be a better understanding of not only the health impact of pacing but also how this impact occurs. This type of interdisciplinary integration is essential for a better understanding of many stressors, not just pacing.

The findings reported in this review demonstrate that there are a host of job features that can negatively influence worker physical and mental health and satisfaction. Much of the evidence is not clear-cut and questions still remain. Yet, we must apply what we currently know to existing problems for which knowledge is adequate. We cannot afford to wait until all of the answers are complete. In answering the questions that remain we must also consider the potential for interactive effects such as the relationship between pacing and boredom or pacing and control. As is often found in the study of toxic chemicals, interactive effects may be synergistic and thus more hazardous. The study of such interactions will demand that more multidisciplinary research be undertaken. Finally, it is imperative that the basic concepts of occupational stress be incorporated into occupational health and safety programmes and educational activities.

References

ARTHUR, R. J., and GUNDERSON, E. K., 1965, Promotion and mental illness in the Navy. *Journal of Occupational Medicine, 7*, 452–456.

BRESLOW, L., and BUELL, P., 1960, Mortality from coronary heart disease and physical activity of work in California. *Journal of Chronic Disease, 11*, 615–626.

CAKIR, A., REUTER, H. V., SCHMUDE, L., and ARMBRUSTER, A., 1978, *Research into the effects of video display workplaces on the physical and psychological function of persons* (Bonn, FRG: Federal Ministry for Work and Social Order).

CAPLAN, R. D., COBB, S., FRENCH, J. R. P., VAN HARRISON, R., and PINNEAU, S. R., 1975, *Job Demands and Worker Health* (Washington, D.C.: US Government Printing Office).

COBB, S., and KASL, S., 1977, *Termination: The Consequences of Job Loss* (Washington, D.C.: US Government Printing Office).

COLLIGAN, M. J., SMITH, M. J., and HURRELL, J. J., Jr, 1977, Occupational incidence rates of mental health disorders. *Journal of Human Stress, 3*, 34–39.

COOPER, C. L., and MARSHALL, J., 1976, Occupational sources of stress: a review of the literature relating to coronary heart disease and mental ill health. *Journal of Occupational Psychology, 49*, 11–28.

COOPER, C. L., and PAYNE, R. (editors), 1978, *Stress at Work* (New York: John Wiley).

FRANKENHAEUSER, M., and GARDELL, B., 1976, Underload and overload in working life: outline and multidisciplinary approach. *Journal of Human Stress, 2*, 35–46.

FRENCH, J. R.P., 1963, The social environment and mental health. *Journal of Social Issues, 19*, 39–56.

FRENCH, J. R. P., and CAPLAN, R. D., 1970, Psychosocial factors in coronary heart disease. *Industrial Medicine, 39*, 383–397.

FRIEDMAN, M., ROSENMAN, R., and CARROLL, V., 1958, Changes in serum cholesterol and blood clotting time in men subjected to cyclic variations of occupational stress. *Circulation, 17*, 852–861.

HARRINGTON, J. M., 1978, *Shift Work and Health: A Critical Review of the Literature* (London: HMSO).

HURRELL, J. J., Jr, 1981, *Psychological, physiological and performance consequences of paced work: an integrative review* (Cincinnati, Ohio: National Institute for Occupational Safety and Health).

KARASEK, R., BAKER, D., MARXER, F., ABLBOM, A., and THEORELL, T., 1981, Job decision latitude, job demands, and cardiovascular disease: a prospective study of Swedish men. *American Journal of Public Health* (in press).

LEVI, L., 1972, *Stress and distress in response to psychosocial stimuli* (New York: Pergamon Press, Inc.).

MARGOLIS, B. L., KROES, W. H., and QUINN, R. P., 1974, Job stress: an unlisted occupational hazard. *Journal of Occupational Medicine,* **16,** 654–661.

MURPHY, L., and HURRELL, J. J., Jr, 1980, Machine pacing and occupational stress. In *New Developments in Occupational Stress,* edited by R. Schwartz (Washington, D.C.: United States Government Printing Office).

QUINN, R. D., and SHEPARD, L. J., 1974, *The 1971-1972 Quality of Employment Survey* (Ann Arbor, Michigan: Survey Research Center).

RUSSEK, H. I., and ZOHMAN, B. L., 1958, Relative significance of heredity, diet, and occupational stress of CHD of young adults. *American Journal of Medical Science,* **235,** 266–275.

RUTENFRANZ, J., COLQUHOUN, W. P., KNAUTH, P., and GHATA, J. N., 1977, Biomedical and psychosocial aspects of shift work. *Scandinavian Journal of Work Environment and Health,* **3,** 165–182.

SMITH, M. J., COHEN, H. H., CLEVELAND, R., and COHEN, A., 1978, Characteristics of successful safety programs. *Journal of Safety Research, 10,* 5–15.

SMITH, M. J., COLLIGAN, M. J., FROCKT, I. J., and TASTO, D. L., 1979, Occupational injury rates among nurses as a function of shift schedule. *Journal of Safety Research,* **11,** 181–187.

SMITH, M. J., COLLIGAN, M. J., and HURRELL, J. J., Jr, 1980, A review of the psychological stress research carried out by NIOSH, 1974 to 1976. In *New Developments in Occupational Stress,* edited by R. Schwartz (Washington, D.C.: United States Government Printing Office).

SMITH, M. J., COHEN, B. G. F., and STAMMERJOHN, L. W., 1981, An investigation of health complaints and job stress in video display operations. *Human Factors,* **23.**

The role of stress in modern industry

Vice President, Labor Relations & Corporate Personnel,
American Telephone & Telegraph Company,
Basking Ridge, NJ 07920, USA

1. Introduction

The subject of stress is drawing the attention of scientists, managers and industrial workers all over the world today. In fact, it has become one of the most important concerns to the social, governmental, business and union leaders.

The subject matter of this book—stress as related to machine-paced work—is certainly not a new topic among researchers. Studies have been going on since as early as the 1930's. Our union associates have been talking with us about aspects of this problem for many years. But if we look at new technology as a change, then occupational stress caused by change—any change—is certainly a contemporary concern among industrial managers; perhaps it is overdue in this day of changing work values and attitudes.

We in the Bell System have certainly become concerned about this phenomenon in recent years. And for good reasons.

The advent of the so-called 'information age' is causing significant changes in the old telecommunications technology. And nowhere is this more evident than in our own business. To keep up with new demands and to provide quality service, we find ourselves implementing new technology at an increasing pace.

But such changes in technology bring with them other changes. They cause changes in work methods, require new skills among workers, demand innovative managerial techniques and often necessitate organizational changes. These events or changes at the workplace have complex relationships among them, and it is not known how exactly they impact on work life. Our own survey results show, however, that they do relate to perceived stress among workers, with possible other impacts on work efficiency and job satisfaction.

I believe the effects just mentioned are not unique to our industry. Many other industries are also experiencing the same problems.

Several of our operating telephone companies have started studies on occupational stress during the past few years. In addition, the Human Resources organization at AT & T is currently involved in an in-depth study of perceived stress and its effects on managers. These limited studies have made one thing quite clear—that we understand very little about this complex phenomenon of stress. For example:

How does one define stress?
What would be a good measure of stress?
How does one clearly differentiate between stress and its symptoms?
How does one measure individual tolerance to stress?
Is all stress bad? The literature tells us that stress can be associated with both positive and negative work conditions.

The answers to these questions are certainly not easy, and we need a better understanding of stress before effective organizational interventions can be undertaken to deal with this problem. However, several efforts are currently under way in our business which we hope will enlighten us further so that we can manage stress more effectively in the future.

2. Research studies

I mentioned earlier that research efforts are currently under way. The primary objective of these studies is to identify the physiological and psychological correlates of stress and to understand how various organizational variables contribute to stress. We have some general understanding but it is not good enough to establish meaningful cause and effect relationships.

3. A system-wide survey

Last year we introduced a Bell System-wide work-relationships survey. Although the overall objective of this survey is to collect information to improve job satisfaction and organizational effectiveness, occupational stress forms a part of this survey. The survey also collects valuable data on employees' perceptions of how change is introduced. Our initial data analyses confirm some of the findings in the literature: that stress is related to certain job characteristics like job autonomy or control, job demand or overload, work methods and interpersonal communication. Tapas Sen in this volume discusses this in more detail.

4. A system-wide QWL effort

About a year ago, The Bell System moved into the area of what is most frequently called Quality of Working Life, although there are many names. To optimize this work, we agreed with Communications Workers of America and other unions to establish national joint union–management committees to work out effective ways of improving both organizational effectiveness and job satisfaction. We believe union–management collaboration is an essential part of any broad approach to improving work quality. Among other things, we hope this effort will encourage and produce worker participation in problem-solving and in other decisions that directly affect their work. The joint commitees have been quite active and are moving towards a more direct role in contributing to the spread of QWL efforts in the field.

5. Technology and human impact

In the past, human factors or what is commonly known as 'man–machine relationships' were the primary areas of investigation in the design of new technology. But recently scientists and engineers in the Bell System are seriously engaged in studying the total human impact of technology. For example, Bell Laboratories task force is studying the psychological and physiological impacts of prolonged use of video display terminals on operators. This study addresses more than biotechnical factors and goes into the dimensions of the work itself which determine satisfaction and motivation. Similar efforts are being made in other Bell Laboratories organizations to understand and estimate total human reactions to new technology from the point of view of stress, performance and job satisfaction. Again, to make this effort more meaningful, we and our unions have established in each company joint union–management committees on the introduction of new technology.

6. Stress management

Several organizations in the Bell System are currently engaged in a few stress management programmes. The objective here is to help those who need help and to find out which methods are effective under which circumstances. Currently these are limited in scope, but we hope that they will evolve as we learn more about effective tools.

7. Conclusion

All of these activities are really an effort on our part to enhance the overall quality of work life in the Bell System. Reduction of job stress is an integrated part of this activity. Although our knowledge is quite limited in this area, we are quite hopeful that our research on stress will enlighten us further about this complex human reaction to modern technology. What is going to make the difference in the future is not a different kind of technology, but the *way* we manage new technology. In this age of competition and economic challenge, managers simply will have to be more innovative about management.

Technology and job pressures

By J. C. CARROLL

Executive Vice President, Communications Workers of America, Washington, D.C.

A review is given of the introduction of new technology by the Bell System during the 1970's and of its effects on various components of the workforce.

1. Introduction

During the 1970's there was a tremendous growth in the implementation of new technology by the Bell System. Computerization changed all aspects of work in the telephone industry. Increasing demands for services and competition have forced the direction of Bell System technology into increasing use of digital methods, computerization and increased versatility of customer equipment.

The net results of technological change on Bell System employees amount to significant changes in occupational distribution, the introduction of new skill requirements, job elimination and job-induced stress. In general, technological change has been very effective in reducing labour intensity, particularly in operator and clerical occupations. For example, the computerized traffic service position system (TSPS) enables long-distance operators to handle twice as many calls. The automation of traditional operator functions has increased efficiency, while effectively reducing the necessity for operator contact during call completion.

The implementation of new technology has also resulted in job reduction and elimination for some occupational groups. As AT & T installs more modular stations throughout the phone system, the concept of the Phone Centre Store begins to have a substantial impact on the number of jobs and their content for service representatives, installers, and repairmen. Similarly, stored-program control systems have resulted in a decline of operator forces.

The effects of technology have also been psychologically detrimental to workers. Rapidly evolving technology and the obsolescing of skills have resulted in boredom, job insecurity, and occupational stress. Moreover, computerization and the way in which it reduces hands-on work with equipment results in worker dissatisfaction and boredom.

2. Methods

Technological change has had and will continue to have a significant effect on jobs within the telephone industry. In some instances, entire jobs have been eliminated; in other instances, the skill requirements have changed dramatically. While some workers benefit from technological change, others, including clerical workers, operators and craftsworkers will be adversely affected.

The central office has been particularly hard-hit by new technologies. Craftworkers are being affected by computerized maintenance and testing systems and changed operating procedures. Increased computerization is exemplified by electronic switching systems (ESS), with their computer-like control capabilities. The speed with which ESS can switch and process calls has revolutionized a number of traditional network operations. In fact, ESS has facilitated the development of other computerized systems which monitor operations and repair the entire phone network. In terms of employment, the ESS requires less than half the non-management personnel, to operate and maintain a traditional switching office. Therefore, the introduction, improvement and expansion of electronic switching systems has had a substantial impact on those jobs which are directly related to the operation, installation and repair of switching systems.

In addition to slashing labour requirements for central office installers and repairers, the skill changes occurring in this area are breaking down traditional occupational divisions. As control functions become more computerized they move into the realm of management. The computer then eliminates many functions of the skilled craft groups, replacing their mechanical skills with familiarity of computer systems and some knowledge of programming concepts.

Line-maintenance operating systems (LMOS) in conjunction with mechanized loop testing (MLT) comprise an automated repair service bureau for testing and troubleshooting customer trouble reports. The development of the automated repair-service bureau and its related computer systems virtually eliminates the position of the test-desk technician, since all operations are computer-controlled.

Women workers, primarily in operator and clerical positions, have historically constituted a major portion of telephone industry employment. However, technological changes in the last two decades have adversely affected the labour-intensive occupations held by women, such as telephone operators.

Stored program control has had a major impact on traffic personnel as it enables the automation of most traditional operator functions and allows new ones in addition. The operator's duties are being changed by an electronic console which automates most of the switching and billing tasks on operator-assisted, long-distance calls. Thus, the operator's role in completing the call has been significantly diminished. Moreover, Bell is currently developing a cellular mobile communications system which will virtually eliminate the mobile operator's function.

Another innovation which is reducing the need for operators is a device which automatically answers intercept (vacant, changed or disconnected numbers) calls. The computer-assembled voice response explains the reason for interruption and gives new number information. The actual impact of these technological developments is two-fold: a considerable increase in operator efficiency and a sizeable reduction in operator forces.

Computers have had a system-wide impact on clerical forces by transferring all information to data bases and eliminating paper handling. Labour requirements are reduced and skills changed to operating computer display terminals. Further, as functions are consolidated, many traditional clerical jobs are eliminated. LMOS, which has computerized customer records and allows instant access to updated records, has reduced clerical tasks by up to 55%.

3. Results

Job elimination, new skill requirements, increased interaction with computers, and job-induced stress are the trends which stand out when considering the occupational effects of new technologies in the Bell System. Ideally, technological changes would make jobs more challenging, requiring higher levels of responsibility, but this has not been the case.

Just as direct long-distance dialing removed the task of call-switching from the operator, labour intensity continues to be reduced by relegating more of the communications process to the customer. Since the advent of modular units, the industry has established phone centre stores where telephones can be bought, sold, or replaced without the intervention of any installation and repair personnel.

Another adverse consequence of technological innovation is job elimination. The toll operator's electronic switching console, which allows the customer to dial collect, credit card, and third-party calls, with limited intervention by an operator, and the computerization of directory-assistance records are concrete examples. Both innovations effectively eliminate traditional job functions and reduce the demand for operators.

Increased computerization has been used to facilitate the over-supervision of work and to expand management control. The most interesting and skilled tasks are relegated to management, while those left to non-management personnel primarily constitute repetitive, machine-paced work. Such reorganization has led to job fragmentation and down-grading.

Expansion of management control over the work is exemplified by the force administration data system (FADS) and a procedure known as 'pricing and loading'. FADS is a computerized system which measures the traffic pattern in a particular office and subsequently estimates force requirements for the next day. Workers receive computer printouts listing their lunch and break times, based on these projections. Pricing and loading is a technique which enables the computer to estimate the amount of time required for a series of switching jobs. The switching technicians receive these schedules from their visual display terminals.

Technological innovation also substitutes worker flexibility for machine-pacing. A prime example of machine-pacing is the TSPS terminal. With the cord board, operators could regulate the pace at which calls were handled. However, with TSPS, calls are routed by an automatic call distributor which means that an operator can handle an unending succession of calls.

It appears that change and technological progress are the common denominators of most occupational stressors. Job-induced psychological stress often results from over-supervision, work speed-ups, fear of job loss, and lack of decision-making within one's job. Other stressful aspects of work include the under-utilization of one's skills and abilities, evaluations, production standards and deadlines. Studies (Caplan *et al.* 1975, Hinkle and Plummer 1952) have also shown that occupational stress results in significant increases in absenteeism and work-related accidents.

4. Discussion

The Bell System is a prime example of how technology reshapes the work-place and changes job content. The impact of the introduction of ESS has affected employees in all job categories from operators to test-desk technicians. When TSPS connects with common channel interoffice signalling, credit card, collect, and bill to third party calls will be completely dialable without operator intervention. Operators are similarly affected by stored-program control systems and their gradual elimination of operator functions. Although craft workers as a whole are being affected by computerized maintenance and testing systems, the central office craft worker will be most acutely affected. The automatic switching control centres effectively centralize monitoring and repair functions, and reduce the number of switching technicians.

With technology-induced speed-ups have come the problem of occupational stress. In addition to the physiological problems resulting from occupational stress, productivity, morale, and the psychological wellbeing of workers are also affected. Other important causes of stress relate to role ambiguity, work overload or underload, lack of decision-making, and job insecurity.

The interrelated issues of job security and job pressures are being addressed by CWA through collective bargaining. General provisions, such as a seniority basis for layoff and slack work periods, and severance pay, act to minimize employee displacement due to technological change. CWA has negotiated an agreement to establish a joint committee on the quality of worklife, giving workers an opportunity to participate in the design and implementation of their own jobs. Job security will be protected through a strong successorship clause and the expansion of the supplemental pay-protection plan. In addition, an advance notice provision was negotiated which also provides for a joint technology change committee. The technology change committee will analyse technological innovations and their implications for workers, prior to the implementation. Another result of the most recent negotiations is the joint occupational job-evaluation committee, which will establish criteria for a comprehensive redefinition of industry job categories. These and other contractual provisions afford our members some protection against the adverse impacts of technological change.

Overall, increased technology has resulted in increased efficiency, work speed-ups, and sizeable reductions in bargaining unit jobs. The idea of improved work life through technological change is based on maintaining skills wherever possible, retraining programmes and increased worker input on the implementation of such changes. Emphasizing the human aspect of work facilitates worker adjustment to technology changes and reduces stress-related problems.

References

CAPLAN, R., CABB, S., FRENCH, J., Jr, *et al.,* 1975, *Job Demands and Worker Health.* US Department of Health, Education, and Welfare Publication No. (NIOSH) 75-169.

HINKLE, L. E., and PLUMMER, N., 1952, Life Stress and Industrial Absenteeism. *Industrial Medicine and Surgery*, Vol. 21.

Section 2.   Models of Human stress

The problem of stress definition

By D. Meister

US Navy Personnel Research and Development Center,
San Diego, California

Stress research presents serious difficulties: lack of definition of the phenomenon being studied; inability to determine whether one is dealing with an independent, dependent or intervening variable or all three simultaneously; inability to differentiate stress from other related phenomena such as fatigue. Some hypotheses about stress dimensions and their research implications are presented.

1. The problem of defining stress

Until and unless terms to describe 'stress' are defined in a reasonably precise manner, research on that phenomenon will be hampered.

There is no satisfactory definition of stress in the literature. What would satisfy one researcher might not satisfy another, but the confusion in the term has been noted many times. (For a representative set of definitions see Gowler and Legge (1980), Burke and Weir (1980), Hamilton (1979), Payne (1979) and Levine *et al.* (1978).) Many books and papers do not even define stress when they discuss it.

Confusion about stress is much more evident in relation to what has been termed 'psychological stress' (of which occupational stress is one form) than to that produced by physical factors such as temperature, noise or acceleration, although the latter too are not without definitional problems. The focus of this paper is therefore on psychological stress.

The difficulty in defining stress is even greater when attempting to differentiate it from related states such as fatigue, 'load' and, even more important, the state of *non-stress*. Some researchers (e.g., Cameron 1974) explicitly define fatigue as a form of stress; others, such as Sharit *et al.* (1979) equate stress with 'psychophysical cost', which implies that it is merely energy expenditure. Stress has even been extended to include job dissatisfaction and boredom (Caplan *et al.* 1975).

If stress is a consistent concept or set of concepts it must possess one or more qualities that differentiate it from other phenomena. If stress is a general reaction to all stimuli, then it is possible that there is nothing distinctive about its research and one need not even use the term.

2. Questions that need answering

If stress is to be defined, certain questions must be answered.

(1) Is the stress response a unitary state, meaning that despite different eliciting conditions, the response to stressors is the same? Or is it a complex of different states? For example, is the stress response to being under enemy fire in combat the same as that experienced in playing an important game or facing a serious surgery? If there are differences in response to these conditions, are the differences qualitative or merely quantitative?

The physiological response to all these situations may be very similar because *bodily mechanisms are capable of only a limited range of variation in response* (at least as compared with cognitive mechanisms). Experientially, however, the three situations (combat, game, surgery) appear to many individuals to be very different.

This variation is mediated by the individual's perception of the stress-inducing situation. Any definition of stress must therefore take the experiential factor into account.

(2) Is stress an independent, dependent or intervening variable? Can we say that certain situations possessing certain characteristics (about which we un-fortunately know little) produce stress whereas others do not?

Since the occurrence of a stress response is highly personal, it can be hypothesized that a stress-inducing situation is stressful only if it is *perceived* as being stressful (except of course in the case of physical stressors, or obviously life-threatening situations). This would seem to dispose of the notion of inherently stressful cognitive stimulus qualities which are necessary if stress is to be an independent variable.

(3) A stressful response may be composed of three elements: physiological (e.g., increased production of catecholamines); performance (e.g., increased error); and experiential (the individual's awareness of his response to the situation, as manifested in tension or anxiety). Does the stress phenomenon require that each element be manifested concurrently? Can stress exist as a physiological or a performance response alone without the individual's experiencing stress? Is it possible to have 'unconscious' stress, such as that referred to in psychoanalytic writings?

(4) One must also ask, how extensively can the term stress be employed? Beginning as a purely physiological phenomenon (Selye's General Adaptation Syndrome (G.A.S.), Selye 1936) it has gradually been extended.

If anything and everything can be a source of stress and if the stress response extends from mild irritation or frustration to profound depression, it is difficult to differentiate stress from other behavioural states. Some theorists assume that if there is an imbalance between an ideal (desired) state and an actual (less than desired) state, stress must exist. For that reason, we have studies in which job dissatisfaction or boredom, for example, are postulated as stress sources (Caplan *et al.* 1975).

One wonders whether a sort of scientific Gresham's law exists in which the excessive generalization of concepts to more and more conditions eventually degrades the utility of the concept. This may have happened to the term 'stress'. It seems unreasonable on a phenomenological basis to equate stress resulting from physical stressors or a life-threatening situation with the stress supposedly caused by frustration. Unless one restricts the situations in which one postulates stress or at least indicates very specifically the conditions under which stress is assumed to exist, the phenomenon becomes somewhat evanescent.

(5) Is stress an abnormal state? Selye's original concept (Selye 1936) postulated arousal as an essentially normal physiological reaction to stimulation; however, as the use of the term has expanded, it has developed a connotation of the undesirable or abnormal. If it is true that stress is undesirable or abnormal, then what is or should be defined as a normal state? To define the stress phenomenon one must differentiate it from a non-stress state, a differentiation that can be made only experientially because physiological and performance indices are too insensitive to differences in stress-inducing situations.

3. Stress dimensions

One way of defining stress is in terms of its dimensions, i.e., ways in which the source of the stress and the individual's response to the stress situation can be described and its variables measured.

The following is for illustrative purposes only.

3.1. *Nature of the stress situation*

Examples are an examination, a football game, surgery or machine-paced work. It is probable that the individual's stress response is strongly influenced by the nature of the situation that elicits it. These can be subdivided into: objectively life-threatening situations; fear-inducing situations that are not objectively life-threatening, e.g., fear of heights; situations that elicit the individual's fear of losing something of value—prestige, money, job; task-accomplishment situations in which the stress (if it is stress) is in the effort to satisfy the requirements of the task. An example of the last is operation of the Post Office's semi-automatic letter-sorting machines. Within this category (which is by far the most common) the source of the pacing may be external (like the speed with which letters are presented to be coded by the letter sorter) or internal (an internal standard of what must be achieved).

3.2. *Stress source specificity*

The stress is either multidetermined as in occupational stress, composed of a number of elements such as inadequate pay, lack of recognition or boredom, which means that the source of the individual's distress will be perceived by him as being relatively non-focused and hence something which he cannot easily overcome, or specific, e.g., a specific job, procedure or individual against which he can take concrete action.

3.3. *Nature of the stress response*

There are experiential and performance responses as well as physiological ones. Experiential responses may vary from: increased effort, possibly in responding to increased task demands; to mild tension; to severe tension; to anxiety/fear. From a performance standpoint, there may be an attempt to escape the stress situation; to attack the source of the stress; or to change one's behaviour to satisfy the stress requirement. All these categories may not be on the same continuum.

3.4. *Response awareness*

This is the degree to which the individual is aware that he is being stressed.

3.5. *Stress consequences*

For example, the consequences of being unable to keep up with a letter-sorting machine are not the same as those involved in anticipating serious surgery or combat.

3.6. *Immediacy of consequences*

This is the speed with which stress consequences are anticipated to occur. An individual who thinks he has only six months to live because of leukaemia reacts differently from someone who, although he has the same disease, thinks he has six years to live.

3.7. *Response availability*

This is the individual's perception of what he can do to modify the stress situation. If he has no options, he is apt to feel more stressed than if he has even one option.

The reason for examining the stress dimensions is that the definition of stress *as it applies to a specific stress-inducing situation* must go beyond general statements like an imbalance between demand and capacity. If one does not know how the stress situation is perceived by the subject, one can hardly interpret his behaviour. These dimensions may also suggest ways of organizing stress research in terms of the variables inherent in them.

4. Stress–response interrelationships

For the author, stress is a dependent variable. Since any situation may induce stress, depending on the individual's perception of that situation, it cannot be an independent variable (except in physical stressors or obviously life-threatening situations). Stress can be defined very grossly as an intervening variable based on theoretical concepts such as Hebb's arousal (1955) or Lazarus's psychological theory (1966), but this is not of much use to us for definitional purposes, because the theories on which such a variable would be based are too vague.

We are therefore on relatively solid ground only in thinking of stress as a dependent variable, manifesting itself in three modes, physiological, perform-ance and experiential. The fact that there are three ways of expressing the stress response means that logically efforts must be made to correlate them or at least to understand their interrelationships. Ideally, the stress state should manifest itself by significant changes in all three modes, but often it does not and then the question is, as was indicated previously, why?

It is possible that the primary manifestation or agent of stress is experiential and that any physiological and performance effects manifested are merely con-comitants; that initially the individual perceives and experiences the stress situation as a cognitive activity and only then are the physiological symptomo-logy and any performance effects triggered—a sort of James-Lange pheno-menon. Obviously, physiological mechanisms are triggered by the stress perception and these in turn trigger the physiological symptomology, but these may be of secondary importance.

5. The need to differentiate stress

The term 'stress' is beguiling and attractive. It must be, because its use has expanded to encompass almost everything. To counter this trend a major aspect of stress research must be *differentiation*. Everything we know suggests that rather than there being a single phenomenon called stress, stress is specific to a set of circumstances involving a task and an individual.

It is necessary, therefore, for researchers to decide whether they are dealing with a ubiquitous phenomenon or one that must be differentiated from other phenomena. If stress is qualitatively and/or quantitatively different from a non-stressful state, it becomes necessary to establish criteria and standards to differentiate the two conditions.

6. Implications

Stress researchers should attempt to define in greater detail the situations they are studying. This is particularly important when one is dealing with 'inferred' stress situations like job dissatisfaction, where the state is deduced from other indices. Because of the importance of the experiential aspect in stress, researchers should develop and *validate* a scale or other means of defining more precisely the subjective experience of stress. Researchers must decide whether stress is associated with every effortful activity or is an abnormal condition. If stress is abnormal, they must develop techniques for discriminating between the stress and non-stress conditions. Researchers should concentrate on 'strong' stress responses, where the stress source is fairly specific and directly experienced. It is possible that only when the stress source is quite marked can it be differentiated from related phenomena such as fatigue. A scheme for categorizing stress situations should be developed so that it is possible to classify and differentiate such situations, particularly if stress is specific to different types of situation.

The opinions expressed are those of the author only and do not reflect those of the Departments of the Navy or Defense.

References

BURKE, R. J., and WEIR, T., 1980, Coping with the stress of managerial occupations. In *Current Concerns in Occupational Stress,* edited by C. L. Cooper and R. Payne (New York: John Wiley).

CAMERON, C. 1974, A theory of fatigue. In *Man Under Stress,* edited by A. T. Welford (London: Taylor & Francis).

CAPLAN, R. D., COBB, S., FRENCH, J. R. P., VAN HARRISON, R., and PINNEAU, S. R., Jr, 1975, *Job Demands and Worker Health, Main Effects and Occupational Differences.* HEW Publication No. (NIOSH) 75–160, Institute for Social Research, University of Michigan Ann Arbor, Michigan (Washington, D.C.: US Department of Health, Education and Welfare).

GOWLER, D., and LEGGE, K., 1980, Evaluative practices as stressors in occupational settings. In *Current Concerns in Occupational Stress,* edited by C. L. Cooper and R. Payne (New York: John Wiley).

HAMILTON, V., 1979, Human stress and cognition: Problems of definition, analysis and integration. In *Human Stress and Cognition: An Information Processing Approach,* edited by V. Hamilton and D. M. Warburton (New York: John Wiley).

HEBB, D. O., 1955, Drives and the CNS (conceptual nervous system). *Psychological Review,* **62,** 243–254.

LAZARUS, R. S., 1966, *Psychological Stress and the Coping Process* (New York: McGraw-Hill).

LEVINE, S., WEINBERG, J., and URSIN, H., 1978, Definition of the coping process and statement of the problem. In *Psychobiology of Stress, a Study of Coping Men,* edited by H. Ursin, E. Baade and S. Levine (New York: Academic Press).

PAYNE, R. L., 1979, Stress and cognition in organizations. In *Human Stress and cognition: An Information Processing Approach,* edited by V. Hamilton and D. M. Warburton (New York: John Wiley).

SELYE, H., 1936, A syndrome produced by diverse nocuous agents. *Nature* (London), **138,** 32.

SHARIT, J., SALVENDY, G., and DEISENROTH, M. P., 1979, Psychophysical effects of paced and self-paced work in relation to motivational, external and internal attentional requirements of a task. *Human Factors Society, Proceedings,* 23rd annual meeting, Boston, Massachusetts.

Organizational stress, cognitive load and mental suffering

By A. WISNER

Laboratory of Work Physiology and Ergonomics, Conservatoire National des Arts et Metiers, 41 rue Gay-Lussac, 75005 Paris, France

The evolution of technology (computerization, automation) connected with classic or recent types of work organization gives rise to situations where activity is not far off being purely cognitive, even in mass-production or low-qualified office work. Many activities, like agriculture or nursing, now have a strong and complex cognitive component. In these conditions, a precise analysis of mental activities at work must be undertaken (perception, identification, decision, short-time memory, programmes of action). This analysis is related not so much to what employees are supposed to do but to what they are really doing.

The signs of mental suffering (complaints, neurotic behaviour, psychosomatic diseases) can then be related to specific aspects of groups of tasks. These aspects characterize especially dangerous types of organization. Among them, it is legitimate to put paced work, but also conflicting situations, current use of multiple codes, frequently interrupted tasks, activities inducing self-acceleration, etc.

1. Introduction

One may sometimes wonder why, of so many factors related to mental suffering, work organization has to be singled out for particular attention. There are several reasons for this:

Salaried employment has become the general rule in our developed societies;

Salaried employment is subject to a work contract under the terms of which work organization is determined by the firm;

The level, stability and quality of production seem easily obtainable through a highly sophisticated and binding organizational set-up;

The time spent at work, the stake of this part of life, the concentration of power in the firm, and the artificiality of today's place of work, are sometimes sources of danger to health; but they equally provide the means of effectively preventing potential problems.

The health risks originating in the organization are related to three main sources of error:

An inaccurate representation of the characteristics of the real population of

workers available, e.g., age, sex, state of health and level of education;
The transformation of legitimate prediction into rules that have to be obeyed; prior work anaysis (the shop floor, the number of machines and the manpower necessary to ensure the production of a new unit) is, of course, necessary but it is dangerous to try to squeeze the production to the prediction, in order to make the economic results as good as those predicted;
The profound ignorance of many engineers and designers of the real characteristics of human physiology and psychology.

The result of these factors is sometimes a rather large gap between what the workers are supposed to do and what they are really doing. When considering possible sources of cognitive overload and mental suffering, we have to refer to real activities: Ergonomic Work Analysis (EWA) is the key to understanding such realities (Wisner 1981).

The main tool of EWA is, of course, behaviour, as well as the verbal description by the worker of what he/she is doing and sometimes his/her image of the functioning of the system (operational image of Ochanine 1971). Physiological measurements are sometimes very useful (Wisner 1971).

If behaviour remains the central object of study, one must examine not only *action behaviour*, which is measured in time and motion studies, but also observation and communication behaviour. *Observation behaviour* is described essentially on the basis of postures and movements of the body, the head and the eyes; for example, number, durations, orientations, sequences of eye fixations and movements in electronic assembling or in correction of texts appearing on VDU screens. *Communication behaviour* is mainly verbal, but it is also semiotic. All verbal expressions during work can be registered on magnetic tape and analysed afterwards from many viewpoints (volume, duration of periods, orientation, cognitive and effective content). Under the term semiotics, one can include not only the formal language of body signals but also the informal expression of the body: an operator may consider that a coworker who has taken off his protective goggles intends to stop welding.

2. The three aspects of work strain

All activities, including work activities, have at least three aspects: physical, cognitive and psychic. Each of them is likely to lead to strain. They are interrelated and it is rather frequent, though it is not necessarily the case, that a high degree of strain in one aspect is accompanied by some degree of strain in the two others. If the definition of the two first aspects are rather obvious, the same is not true for the psychic dimension, which may be defined in terms of the level of conflict between the conscious or unconscious representation of the relations between the self and the situation (in this case work organization). But it is also the level where physical suffering and fatigue, lack of sleep induced by the nycthemeral distribution of work periods, and cognitive overload may induce some alteration of mood.

The comments presented here will be related to the situations where cognitive overload is predominant. But one must remember that the three aspects are always present. For example, the activity of a man who delivers goods to city grocers seems to be mainly physical. Many ergonomic studies have considered this aspect and produced acceptable results. But the cognitive dimension should not be neglected, for it may sometimes be predominant: choice of itinerary, numbering of bottles, control of bills and sometimes of money. The psychic aspect of the task is sometimes hidden but at other times is predominant: aggressive attitude of the grocers towards delays, changes of price, difficulties with car drivers during obstructive parking in front of the grocer's. These psychic dimensions leading to mental suffering may explain rapid personnel turnover.

At the other extreme, the work strain of receptionists may be considered as purely psychic, especially in some offices where these employees are there to receive the legitimate or illegitimate protests of the public against the organization. In fact, social workers doing their job correctly have frequently a high cognitive load in view of the difficulty of understanding the demands of people who are sometimes quite unaware of the administrative jargon. Some of these situations may also possess some severe physical dimensions if the task requires distributing heavy documents or goods, or accompanying people to different parts of a big building.

3. Tasks with predominant cognitive strain

Though the tasks with predominant cognitive strain have existed for a long time (telephonists, accountants, teachers), their number is growing quickly, mainly through computerization.

The situations considered will be those where the task is strictly organized and where pace produces some speed stress. It should be remembered that high mental strain may also be found in complex situations with many interfering tasks (Theureau 1979) and a stress proceeding from the disproportion between demand and personnel resources (nurses, educators, sales persons).

Perceptive difficulties should not be underestimated, since they increase the mental effort required and, possibly, the anxiety which comes from uncertainty of comprehension. Messages, whether verbal or non-verbal, transmitted orally or by acoustic means, may be deformed or partly drowned out. The problem of orally transmitted messages is particularly acute when the recipient is not very familiar with the language of the speaker (such as a foreign worker listening to his superior speaking against a noisy background) (Rostolland 1979). Similarly, vibrations may make it difficult to read the displays on read-out equipment. However, perceptive difficulties at work are due for the most part to lighting problems or to the visual characteristics of the job. Let us take as an example the operation of computer terminals (visual display units). Some researchers (Grandjean 1980, Meyer *et al.* 1978) have even gone so far as to assert that visual disorders observed among terminal operators can mostly be blamed on poor screen quality, on the characters displayed (flickering, blurred edges) and on the lighting (reflections on the protective glass covering the screen). Similarly, in the textile and electronics industries, difficulties of perception contribute to the mental effort required in carrying out the tasks concerned.

As regards the *cognitive content of the task itself*, the foremost aspect is decision-making, no matter how minor the decision (for example, the decision to fit resistance F35 at point H17 on the mounting plate during assembly of electronic apparatus). The maximum decision-making capacity of the human brain is very low (ranging from 15 bits/min under stable conditions to 50 bits/min at peak effort). Beyond this rate the brain is overtaxed if the only cognitive activities are of the decision-making type (Kalsbeek 1968).

However, decision-making is far from being the only element in the cognitive activity, nor is it the principal one. The difficulties of perception will be recalled, and to these must be added problems of identification and recognition. The most critical element is probably memory, whether immediate, i.e., requiring mental effort throughout the memorizing period (this is 'active' memory as compared with the passive memory of a computer), or long-term, particularly retrieval of information. Memory capacity is known to be low when the individual is tired and especially when he is suffering from lack of sleep. But, conversely, high cognitive stress preceding the night rest period induces sleeping difficulties (Vladis and Foret 1981).

It has frequently been found that workers engaged in predominantly mental tasks complain of *physical distress*, such as pains in the back and neck and visual disorders (stinging and burning sensations in the eyes, double vision, etc). These disorders are generally attributed to a high degree of immobility combined with sustained concentration. It has been shown (Laville 1968) that among female workers in the electronics industry the rigidity of posture increased with the difficulty and speed of the task, and the head was held closer to the workpiece. Under laboratory conditions, the above-mentioned researchers demonstrated from EMG traces that the activity of the neck muscles increased with the frequency and complexity of the signals being observed by the workers. Similarly, Duraffourg *et al.* (1979) showed that among operators using VDUs the number of visual fixations is proportional to the density of the information contained in the text, while the duration of the fixations is proportional to the difficulty of the codes employed.

Thus, the need to observe and 'process' signals leads to postural immobility, where the eyes are brought close to the work and the postural muscles contract excessively; consequently, pain is felt in the back and neck. Moreover, staring attentively at a difficult piece of work causes fatigue of the intrinsic muscles (accommodation) and of the extrinsic muscles (convergence) of the eye, as well as irritation of the conjunctiva due to desiccation (brought about by insufficient blinking).

4. Job content and mental suffering

Thirty years ago, a study by Le Guillant (1952) showed the extent of the cognitive demands placed upon female telephone operators and the fair degree of uniformity in their reactions to the constraint of the job. The "telephonists' neurosis" described in the study consisted of headaches, buzzing and whistling noises in the ears, obsessive thinking, stereotyped speech patterns, sleep problems and moodiness. These disorders were in evidence not only at work and during subsequent rest periods, but also during days off and at the beginning of holidays.

The expression "telephonists' neurosis" may be debatable, and the term 'neurotic syndrome among telephonists' preferable, since the job does not create the neurosis but merely reveals its outward expression. It should also be noted that some operators were either not affected at all or only affected to a slight degree, whereas others were quite unable to stay on the job.

Since that time, it has been shown that this neurotic syndrome occurred in all jobs requiring a high degree of mental effort (key-punch operators, workers in the electronics and textile industries, VDU operators). The only variations lie in the outward manifestations, which are specific to the particular constraints of the job. Instead of the auditive problems noted among telephonists, one finds back and neck pains among electronics and textile workers and visual and para-vertebral problems among VDU operators. But the basic fact remains that workers subjected to a major mental effort suffer from a neurotic syndrome.

The Le Guillant syndrome has complex roots: high cognitive speed stress, ambiguity of the task, difficult relations with the public.

The relations between mental suffering and cognitive speed stress has been clearly demonstrated experimentally by Kalsbeek (1968) who considered only one aspect of cognitive strain, microdecisions. The subjects were confronted with a two-fold task. The main task consisted in stepping on the left-hand pedal when a green signal light was turned on, and the right-hand pedal when a red light was turned on. Signals were flashed haphazardly. The secondary task consisted in writing a text of the subject's choice. With increasing signal frequency the texts, which at first were interesting, became puerile before degenerating into a repetition of words, then letters, and finally into an illegible scribble. When the signal rhythm slowed, the process was reversed. When the experiment lasted for some time then stopped, the subject became aggressive. Sometimes he was disoriented and would head for a wall instead of the door when leaving a room with which he was familiar.

The short, intensive experiments are too close to the realities observed every day in the place of work not to be extremely significant. We are also familiar, in mass-production workshops, with emotional outbursts such as an attack of nerves or fainting in workshops staffed by women, and the 'slanging match' or violent gestures directed against the machinery in workshops employing men (parenthetically, one will note the difference between affective expressions of emotion socially 'authorized' according to sex). More specifically, it will be noted that such emotive crises usually occur during the learning period. It is a fact that the time alloted for job familiarization is as a rule much too short, and this period is accompanied by marked overwork. Experienced supervisors know that when such crises occur, some workers break down and quit their jobs whereas others who have overcome this barrier continue for a long time.

The memory of these crucial periods is so painful that it determines later attitudes. A study on workers in nine French electronic firms found that it was precisely those female staff who found their job particularly hard that did not want to change. The reason was that they were afraid of another familiarization phase after having experienced the previous one. Clearly, the resistance to change may be based on quite objective grounds.

Ambiguity inside the task is a frequent occurrence. For example, in a spectacle-lens manufacturing plant, workers reprimanded in connection with quality control became anxious since they failed to understand the criteria determining lens rejection. The solution to this anxiety was found simply by inserting in the production line, after every 20 lenses, a 'standard' lens displaying maximum-admissible defects.

No discussion of this subject would be complete without referring to Pavlov's pioneering experiments. As is well known, dogs stimulated alternately with a pleasure-linked signal (food) and a pain-linked signal (electric shock) became neurotic when the signals became so similar as to be indistinguishable. The neurotic disorders took the form either of aggressiveness or of falling asleep. After a time, some dogs began to suffer from psychosomatic disorders (ulcers of the digestive tract). It should also be stressed that some dogs seemed to stand up to the tests better than others; in any case, they reacted differently.

A particularly high rate of absenteeism, due mainly to nervous breakdown, has been observed in situations where *contact with the public* is involved. The most pathogenic work situations are those combining a heavy workload (e.g., which may take the form of long queues of people) and a distressed attitude on the part of the public concerned (employment departments, claims departments and—once again—telephone switchboards).

As if to protect the personnel from the pressure of the public, barriers have progressively been erected. These barriers may be physical (windows perforated with holes) or organizational (individuals take a numbered ticket on arrival and are called in numerical order, or access to the counter or window is restricted, etc.).

Obviously, under these circumstances very special relationships are established which are analogous to the phenomenon of aggressive transfer. This is a very subtle social process in which decisions are taken remote from the public, while clerical staff are given the job of dealing with a dissatisfied public (although such staff may in fact be competent, they are in any event put there to bear the brunt of public expressions of disappointment or cantankerous incomprehension).

To treat this problem at an individual or technical level seems quite inadequate, for such situations are the perverse outcome of social organization.

5. Conclusion

One of the most remarkable characteristics of living creatures is the diversity of their reactions to a given situation. Among any population, reactions will vary widely to the same alcohol content, or to the same exposure to benzol or to noise. Even Pavlov's dogs reacted differently to the same conflict situation. One can therefore expect a wide range of tolerance to work situations.

A person who takes on a job brings with him his genetic capital, the sum total of his pathological history dating back through birth to his existence 'in utero', and the marks of his accumulated physical and mental experiences. He also brings with him his living habits, customs and learning experience, which together influence the personal cost to the individual of the work situation in which he is involved.

To go back to our main theme of mental workload and suffering, the problems arise from the relationship between history of the individual and history of the society following M. Plon's views. More specifically, Dejours (1980) discussed the difficult, even impossible, relationships between the individual, and his need for 'pleasure' on the one hand, and the 'organization', tending towards total constraint and conformity with a machine-like model, on the other. Such are the roots of the problem. However, many particular aspects of organization, some of which have been described, are particularly and intolerably constricting. They trigger dangerous reactions specific to different people. It is important, therefore, to be aware of these reactions and to avoid them in designing production systems.

References

DEJOURS, C., 1980, *Travail: Usure mentale* (Paris: Centurion).
DURAFFOURG, J., GUERIN, F., JANKOVSKY, F., PAVARD, B., 1979, Analyse des activités de saisie-correction des données dans l'industrie de la presse. *Travail Humain*, **42** (2), 231–243.

GRANDJEAN, E., 1980, Ergonomics of VDUs: review of present knowledge. In *Ergonomic Aspects of Visual Display Terminals,* edited by E. Grandjean and E. Vigliani (London: Taylor & Francis).

KALSBEEK, J. W. H., 1968, Measurement of mental work load and of acceptable load. *International Journal of Production Research,* **7**, 33–45.

LAVILLE, A., 1968, Cadences de travail et posture. *Travail Humain,* **31** (3.4), 73–94.

LE GUILLANT, L., 1952, La psychologie du travail. *La Raison,* **4**, 75–103.

MEYER, J. J., REY, P., KOROL, S., GRAMONI, R., 1978, La fatigue oculaire engendrée par le travail sur écran de visualisation. *Sozial und Präventiv Medizin,* **23**, 295–296.

OCHANINE, D., 1971, L'image opérative effectrice. *Questions de psychologie,* **3** (published in Russian, translated into French).

ROSTOLLAND, D., 1979, *Contribution à l'étude de l'audition de la parole en présence de bruit: caractéristiques physiques, structure phonétique et intelligibilité de la voix criée.* Thèse Doctorat d'Etat, Paris.

THEUREAU, J., 1979, *L'analyse des activités des infirmières des unités de soins hospitalières.* Thèse Docteur Ingénieur CNAM, Paris.

VLADIS, A., FORET, J., 1981, Effets de la charge cognitive de travail sur le sommeil consécutif, à paraître, in *Travail Humain.*

WISNER, A., 1971, Electrophysiological measures for tasks of low energy expenditure. In *Measurement of Man at Work,* edited by W. T. Singleton, J. G. Fox and D. Whitfield (London: Taylor & Francis) pp. 61–72.

WISNER, A., 1981, Méthodologie ergonomique. In *Physiologie du Travail-Ergonomie,* edited by J. Scherrer (Paris: Masson).

Job decision latitude, job design, and coronary heart disease

By R. A. Karasek

Department of Industrial Engineering and Operations Research, Columbia University, USA

Job-design strategies in industrial society have emphasized restriction in skill requirements and decision authority for major groups of workers, to enhance productivity. This principle is questioned as to its productive efficiency. Furthermore, low skill requirements and decision authority are shown to be related to psychological strain, at the individual and occupational level, and to the prevalence of cardiovascular illness in the USA, and to both prevalence and incidence of cardiovascular illness in Sweden.

1. Introduction

A vast range of factors must be examined to account for the effects of the psychosocial work environment on psychosomatic health, on individual motivation, and job-developed behaviour patterns. However, it appears that a significant level of understanding can be achieved by beginning with the joint interaction of just two major job dimensions which have substantial significance for both organizational design and for occupational health. These two job dimensions are job decision latitude and workload demands (figure 1). The joint model based on the interaction of job decision latitude with workload demands was originally developed to predict psychological strain from working conditions (Karasek 1976, 1979a). Obviously, these character-

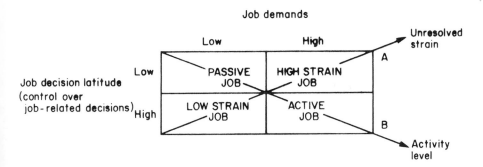

Figure 1. Interaction of job decision latitude and workload demands.

istics of work are also of substantial importance for production organization. Individual workload is closely linked to individual performance quotas and firm output level, while individual decision-making opportunities are a clear result of organizational structure and division of labour.

Strategies for productivity improvement have focused primarily on changing the decision structure and the division of labour. In particular, these strategies have focused on specific changes in two dimensions of decision-making: *intellectual discretion* (level of skill of usage) and *decision authority*. In US studies concentration has usually been on decision-making opportunities at the task levels, e.g., time-scheduling of tasks, machine-pacing, opportunities to affect the whole product, decisions about who to work with and how to subdivide the task, and measures of supervisory closeness. Commonly used measures are Kohn and Schooler's (1973) well known framework for 'occupational self-direction', consisting of 'substantive complexity' of the work (and its 'repetitiveness'), and 'closeness of supervision'. Hackman and Lawler's (1971) Turner and Lawrence-inspired (1965) Motivating Potential Score has as its core measures of 'variety' (*range* of skills which may be utilized) and job 'autonomy'. Gardell's (1971) scale of task control includes a 'qualification' subscale and a 'freedom' subscale. The US Dictionary of Occupational Titles rates skill usage on three scales (things, people, data), but also includes measures that assess task-level decision-making authority (Spenner 1980). Finally, research by Karasek on decision latitude has identified two subcomponents, 'intellectual discretion' and 'decision authority' which are rather highly correlated ($r = 0 \cdot 48$; US Quality of Working Life Survey) but which still may vary independently in some jobs. Both of these job decision latitude dimensions have been reduced in many job-design solutions under Taylor's scientific management philosophy for blue-collar workers.

2. Job-design strategies, restricted decision opportunities, and productivity

Adam Smith (1776) recommended specialization of labour for productivity enhancement. This technique would increase the 'skillfulness' of each worker by focusing his efforts on one simple, repetitive task (Kilbridge 1966). Unfortunately, modern applications of this technique are often focused first on a narrowing of the *range* of skills—a policy that was never advocated by Smith. The implicit presumption is that there are strict limits to the worker's 'learning capacity' and that an effective way to assure *depth of skill* in one area was to restrict the overall *range* of skills in other areas to the absolute minimum. However, if overall learning capacity is less limited, workers may be able to develop sufficient depths of skill in a wider range of activities. Related managerial techniques were recommended by Fredrick Taylor (1947). His underlying supposition was that the mass of the existing workforce could not attain skill levels or judgemental sophistication sufficient to justify significant decision authority over task organization and co-ordination. However, the level of formal training of the workforce has changed drastically since Taylor's time (i.e., the portion of the workforce that completed high school in 1890 was 3·5%, as opposed to 60% in 1970).

Now the more salient problem in Western industrial societies is under-utilization of skills in highly educated populations. While this has come to be known as the 'overeducation problem' by some economists (Freeman 1976), in fact it may well be the result of 70 years of devoted application of the principle that jobs had to be 'de-skilled' to correspond to an outdated presumption of minimum intellectual capabilities. In any case, this now represents an under-utilization of a major national productivity resource which, if not redressed by substantial transformation of job-design theory, could drastically hinder productivity growth in the USA (Karasek 1979b). Table 1 shows that the actual educational level of the USA population has increased continually over time, but for the first time the skill level requirements of jobs in the USA economy declined between 1960 and 1970. This occurred in spite of the apparently increasing technological complexity of USA industry. During the same period the quadrennial US Quality of Employment Survey, based on detailed interviews of a random sample of USA workers, found that the percentage of workers claiming significant skills which they were not able to use increased from 27% in 1969 to 36% in 1973 (Quinn and Staines 1979).

Table 1. USA workforce compared with respect to the average general educational development requirement (general skill level) of jobs and the level of actual education (Berg *et al.* 1978).

	1950	1960	1970
Required skills (G.E.D.)			
8 years or less	33%	9%	8%
8 years to 15 years	59%	73%	77%
15 + years	8%	18%	14%
Actual education			
11 years	61%	53%	39%
12–15 years	32%	38%	48%
16 + years	9%	9%	13%

These findings are consistent with micro-level findings from a five-plant survey (Karasek 1981, Camman *et al.* 1974). Figure 2 shows that much higher levels of skill under-utilization occur for workers at lower levels of job decision latitude. Indeed, there is *over*-utilization of workers at high job decision latitude levels. This suggests that low-decision latitude jobs are much lower in skill utilization than is necessary to fit the capabilities of workers and that

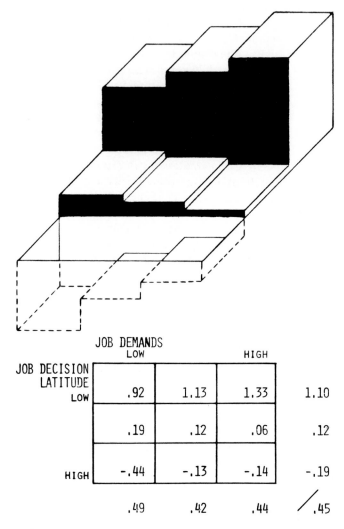

(SKILL POSSESSED -- SKILL REQUIRED)
MEASURED BY EDUCATIONAL LEVEL: 0 YRS.= 1, 17+ YRS.= 7
UNIVERSITY OF MICHIGAN QWL EFFECTIVENESS SURVEY (N=222)

JOB DEMANDS

	LOW		HIGH	
JOB DECISION LATITUDE LOW	.92	1.13	1.33	1.10
	.19	.12	.06	.12
HIGH	-.44	-.13	-.14	-.19
	.49	.42	.44	.45

Figure 2. Skill under-utilization by job characteristics.

there is an exaggerated inequality of skill usage and control opportunities in the workforce that cannot be justified by individual differences in capabilities.

A review of the distribution of job decision authority suggests that under-utilization may also occur with respect to 'judgemental capacities' of the USA workforce. Substantial discrepancies may exist between relative skill levels and decision authority. Frankenhauser and Gardell's (1976) lumber graders at an automated mill are an example of high skill with little authority over job organization. Another simple example in our own research with the US Quality of Working Life Surveys are typesetters. There are also significant examples of a skill–authority discrepancy during early career stages. In figure 3 we observe relative skill level required versus decision authority (units are standard deviations of the national workforce on the respective dimensions) for three broad occupational groups. We can see that for high-status white-collar workers (i.e., professionals, managers, sales personnel) skill level and authority are high from early career and remain high and closely matched for older population segments (age cross-sections in ten-year increments). However, for highly skilled blue-collar workers such as craftsmen, repairmen and other skilled trades, relative authority lags substantially behind relative skill levels. 'Judgemental under-utilization' is high. A similar judgemental under-utilization occurs even for low-skilled blue-collar workers. This is a group that simultaneously has the capacity to handle the higher-skilled jobs (table 1), making potential 'judgemental under-utilization' even larger. Thus, among the younger half of the USA workforce, many groups of blue-collar

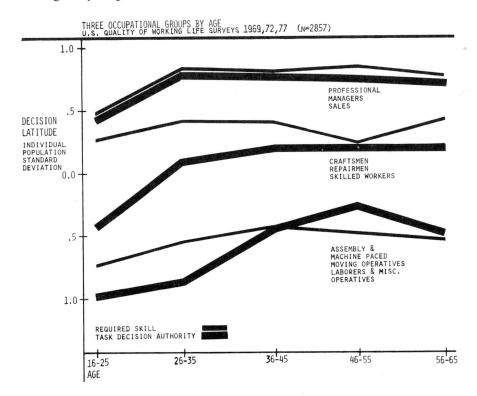

Figure 3. Required skill level versus task decision authority.

workers have substantially under-utilized skill and judgemental capacities. This does not appear to be true for high-status white-collar workers.

3. Job dimensions and psychosomatic illness

Job-design strategies limiting skill breadth and decision authority may also contribute to psychosomatic illness. Such consequences may be analysed by simultaneously examining the effects of other job dimensions such as workload, job insecurity and social support along with decision latitude. Figure 1 summarizes in a simple way our model of the types of job that might result from different combinations of job demands and job decision latitude (Karasek 1979a, Karasek *et al.* 1981). Two interactions are actually represented by the labelled diagonals, one representing *disproportional* levels of job demands and job decision latitude (A), and the second representing the situation where they are *matched* (B). It is in the first situation, when demands are relatively greater than decision latitude, where we predict that mental strain will develop (Turner 1980, Freeman and Jucker 1980, Karasek 1979a, Gardell 1971, Goiten and Seashore 1980, Aronssen 1980, Detzer 1980). In the second situation, the match of the demands and permitted decision latitude is likely to result in the optimal learning (or 'unlearning') situations (Karasek 1976, 1979a, 1981, Goiten and Seashore 1980).

4. Relating job characteristics, psychological stress and coronary heart disease

For our USA findings we will compare job characteristics for each occupation with coronary heart disease (clinically measured) using data which is available from the US Center for Health Statistics (1960–62 Health Examination Survey $n = 1944$ employed males; 1971–75 Health and Nutrition Examination Survey $n = 2157$ employed males). Our system for job characteristics for each occupation is based on US Department of Labor Surveys of the full USA workforce in 1969, 1972 and 1977. These surveys respectively contain 993, 1292 and 1361 males and 540, 470 and 547 females interviewed in a nationally representative household survey of the working population over the age of 18, working more than 20 hours per week. These surveys have been reported on extensively in other government documents (Quinn and Staines 1979) and in studies by this author (Karasek 1979a) and other research groups.

Related research has shown that the highest concentration of psychological strain (as measured by a composite of sleep problems, physical stress symptoms, depressed mood and job dissatisfaction) is clearly in the high-strain quadrant of figure 1. These associations between job characteristics and psychological strain are not likely to be strongly influenced by self-report biases or selection since they are based upon occupationally aggregated data. These associations between psychological strain and low decision latitude/high demands is also confirmed at the individual level with USA and Swedish national data (figure 4 from Karasek 1979a).

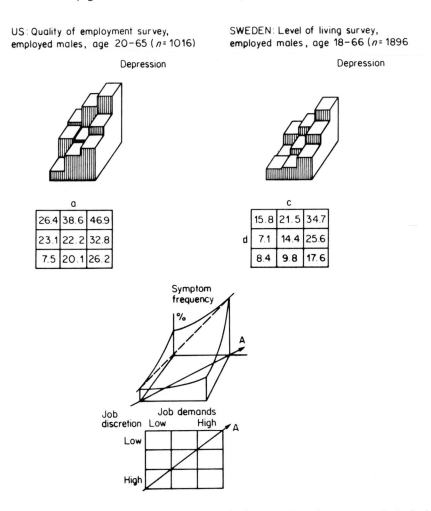

Figure 4. Confirmation at the individual level of associations between psychological strain and low decision latitude.

Associations between job characteristics and CHD have not been heavily investigated and most of the literature to date contends that the findings are not entirely clear, particularly when certain job risk factors are to be isolated (see Karasek *et al.* 1981 for discussion). We feel that much of this ambiguity

R. Karasek

can be traced to an incomplete model of job stress. The US Health Examination Survey from 1960–62 and the US Health and Nutrition Examination Surveys from 1971–75 allow examination of occupations and clinically assessed coronary heart disease. Figure 5 shows the distribution of definite and

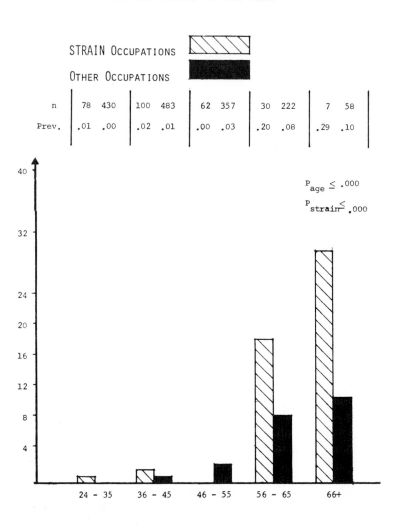

H.E.S. STUDY 1961-1962
PREVALENCE OF DEFINITE M.I. AND ANGINA
AMONG MEN (WHITE) BY AGE GROUP

STRAIN OCCUPATIONS

OTHER OCCUPATIONS

n	78	430	100	483	62	357	30	222	7	58
Prev.	.01	.00	.02	.01	.00	.03	.20	.08	.29	.10

$P_{age} \leq .000$

$P_{strain} \leq .000$

24 - 35   36 - 45   46 - 55   56 - 65   66+

Figure 5. Distribution of definite and suspected myocardial infarction and angina pectoris for high-strain occupations compared to all other jobs by age decades for the USA white male workforce.

CORONARY HEART DISEASE INDICATOR* CROSS-SECTIONAL PREVALENCE IN 1974 AND PROSPECTIVE PREVALENCE IN 1974 AMONG ASYMPTOMATIC RESPONDENTS IN 1968 BY JOB CHARACTERISTICS

(THE PROPORTION OF EMPLOYED MALES WITH THE INDICATION OF CHD DISPLAYED BY JOB CHARACTERISTICS: JOB DEMANDS AND INTELLECTUAL DISCRETION)

CROSS-SECTIONAL PREVALENCE IN 1974   "PROSPECTIVE" PREVALENCE IN 1974

JOB DEMANDS

	LOW		HIGH	
LOW	.032 (63)	.128 (78)	.200 (30)	.105
INTELLECTUAL DISCRETION	.068 (205)	.066 (287)	.104 (144)	.076
	.044 (114)	.040 (202)	.045 (222)	.043
HIGH	.000 (40)	.022 (93)	.028 (143)	.022
	.050	.059	.065	.059 (N = 1621)

JOB DEMANDS

	LOW		HIGH	
LOW	.018 (57)	.069 (116)	.050 (40)	.052
INTELLECTUAL DISCRETION	.050 (219)	.081 (335)	.099 (111)	.074
	.000 (70)	.016 (123)	.045 (111)	.023
HIGH	.000 (36)	.000 (88)	.038 (155)	.022
	.031	.056	.055	.056 (N = 1461)

* ACHE IN BREAST, DYSPNEA, HYPERTENSION, HEART WEAKNESS

Figure 6. Associated between job decision latitude, psychological job demands, and the CHD-incidence indicator for Swedish male workers.

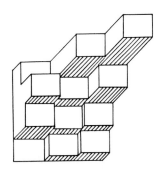

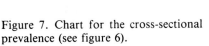

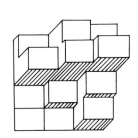

Figure 7. Chart for the cross-sectional prevalence (see figure 6).

Figure 8. Chart for the prospective prevalence (see figure 6).

suspected myocardial infarction and angina pectoris for high-strain occupations compared to all other jobs by age decades for the USA white male workforce. (Similar associations were found in the USA HANES data

1971–75, see Karasek *et al.* 1980). The statistically significant associations also show that the work populations in the high-strain jobs diminished significantly with age, suggesting that many of the potential victims of such work environments have switched to jobs of other types, probably weakening the observable job characteristics/CHD associations.

Analysis of Swedish data on CHD and job characteristics shows consistent results. The Swedish data on job characteristics is available at the individual level and the data base is longitudinal, allowing a much more accurate assessment of causal relationships. The coronary heart disease measure in this case is a self-report measure of CHD symptoms with a $5 \cdot 0 : 1$ risk of CHD mortality in a six-year CHD mortality study (Karasek *et al.* 1981). Figure 6 shows the association between job decision latitude, psychological job demands, and the CHD-incidence indicator. The chart for the cross-sectional prevalence is shown in figure 7, and that for the prospective prevalence in figure 8. The associations, when tested in a multivariate logistic regression analysis simultaneously with smoking, education, age and overweight, show significant relationships in the predicted direction: low decision latitude is associated with increased risk of heart disease, as are high psychological demands. The associations are confirmed in a case-control study of CHD mortality and individual level job-characteristic data.

Job-design strategies advocating limited skill usage and decision authority for the majority of the workforce appear to be associated with a host of undesirable, unintended consequences ranging from skill under-utilization (and consequent productivity loss) to increased risk of coronary heart disease.

References

ARONSSEN, G., 1980, *Lokaltrafiken Som Artetsmiljo Delrapport 1; Arbetsforhallanden Och Halsa*. Preliminary Report, Psychological Institute, Stockholm University.

CAMMAN, C., QUINN, R., BEEHR, T., and GUPTA, N., 1974, *Effectiveness in Work Roles Report 1: Validating Quality of Employment Indicator*. Mimeo, Institute for Social Research, University of Michigan.

DETZER, M. M., 1980, *Presentation on Job Demands in the Bell System to the AT and T Bargaining Committee*. Submitted by the T.I.U. Bargaining Committee.

FRANKENHAUSER, M., and GARDELL, B., 1976, Underload and Overload in Working Life: Outline of a Multidisciplinary Approach. *Journal of Human Stress*, **2**, 35–46.

FREEMAN, H. J., and JUCKER, J., 1980, *Comparing Traditional and Innovative Production Organizations*, Department of Industrial Engineering, Stanford University. Paper presented at the Conference on Current Issues in Productivity, Columbia University, April 1980.

FREEMAN, R., 1976, *The Overeducated American* (New York: Academic Press).

GARDELL, B., 1971, *Produktionstnik Och Arbetsgadje* (Stockholm: Personaladministrativa Radet).

GOITEN, B., and SEASHORE, S., 1980, *Worker Participation: A National Survey Report*, Survey Research Center, University of Michigan.

HACKMAN, J., and LAWLER, E., 1971, Employee Reactions to Job Characteristics. *Journal of Applied Psychology Monograph*, **55**, 259–286.

KARASEK, R., 1976, *The Impact of the Work Environment on Life Outside the Job.* Ph.D. Thesis, Massachussetts Institute of Technology. Distributed by U.S.N.T.I.S.: DLMA 91-25-75-17-1.

KARASEK, R., 1979a, Job Demands, Job Decision Latitude, and Mental Strain: Implications for Job Redesign. *Administrative Science Quarterly*, **24**, 285–306.

KARASEK, R., 1979b, *A New Model of Job Characteristics and Productivity. The Interactive Effect of Skill Utilization and Work Load on Job Performance and Job Stress.* Paper presented at the Conference on Current Issues in Productivity, Columbia University.

KARASEK, R., 1981, Job Socialization and Job Strain: The Implications of Two Related Mechanisms for Job Design. In *Working Life*, edited by Gardell and Johannson (Chichester: John Wiley).

KARASEK, R., SCHWARTZ, J., THEORELL, T., PIEPER, C., and SCHNALL, P., 1980, *Job Conditions, Occupation and Coronary Heart Disease*, Research sponsored in part by the National Institute for Occupational Safety and Health, Grant no. 5 RO1 OH00906-02.

KARASEK, R., BAKER, D., MARXER, F., AHLBOM, A., and THEORELL, T., 1981, Job Decision Latitude, Job Demands and Cardiovascular Disease: A Prospective Study of Swedish Men. *American Journal of Public Health,* **71** (7), 694–705.

KILBRIDGE, M., and WEBSTER, L., 1966, An Economic Model for the Division of Labor. *Management Science*, **12**, 6.

KOHN, M., and SCHOOLER, C., 1973, Occupational Experience and Psychological Functioning, An Assessment of Reciprocal Effects. *American Sociological Review*, **38**, 97–118.

QUINN, R., and STAINES, G., 1979, *The 1977 Quality of Employment Survey.* Survey Research Center, Institute for Social Research, the University of Michigan.

SMITH, A., 1977, *An Inquiry into the Nature and Causes of Wealth of Nations*, edited by E. Cannan (Chicago: University of Chigago Press).

SPENNER, K., 1980, Occupational Characteristics and Classification Systems: New Uses of the Dictionary of Occupational Titles in Social Research. *Sociological Methods Research*, **9**, 2, 239–264.

TAYLOR, F., 1947, *Scientific Management* (New York: Harper Brothers).

TURNER, A., and LAWRENCE, P., 1965, *Industrial Jobs and the Worker* (Boston: Harvard).

TURNER, J., 1980, *Computers in Bank Clerical Functions: Implications for Productivity and the Quality of Life.* Ph.D Thesis, Columbia University, Department of Industrial Engineering and Operations Research.

Stress and human performance: a working model and some applications

By A. F. SANDERS

Institute for Perception TNO, Soesterberg, The Netherlands
and
Tilburg University, Tilburg, The Netherlands

An outline is presented of a framework aiming to relate energetical and
structural mechanisms of information processing and to incorporate a
cognitive stress concept in human performance research. The structural
part is based upon linear stage models following the additive factor logic,
while the energetical part stems from the Pribram and McGuinness theory
about the control of attention. Connections between structural and
energetical mechanisms are postulated as well as the operation of cognitive
evaluation mechanisms, supervising the quality of performance in relation
to the actual demands. Some consequences of the framework are discussed
and predictions for specific performance studies are formulated.

1. Introduction

In human-performance theory 'stress' is usually treated as a collective noun
for various environmental and motivational conditions affecting performance
(Broadbent 1971, Hockey 1979). In contrast, the view is rapidly gaining acceptance that stress is an intervening variable, arising when the organism perceives
unacceptable deviations from optimum conditions which are not easy to
restore (Welford 1973). Or, in other words, a stress reaction occurs on the basis
of negative cognitive appraisal of the situation, rather than of the situation *per
se* (Lazarus 1967). This has the implication that, within the context of human-
performance theory, research about stress does not coincide with research
about effects of sub-optimal performance conditions but, rather, the former
constitutes a subset of the latter. The aim of this paper is to present an outline
of a working model for research on stress as well as on sub-optimal conditions,
to depict some consequences and to suggest certain experimental tests.

2. Stress and arousal

The most simple notion of stress, which has been also prevalent in the
explanation of environmental effects, concerns a description in terms of
arousal. Starting from the well known Yerkes–Dodson curve, relating arousal
and performance, stress would arise at an unacceptable sub-optimality, both

in the case of over- and under-arousal. The larger the deviation, the stronger the stress reaction, in terms of a physiological or hormonal response pattern (Levi 1971).

The main assumptions underlying this extremely simple scheme concerned unidimensionality and aspecificity of arousal as well as of stress. Both assumptions were popular in the fifties (Hebb 1955, Selye 1956) and well in accord with the classical sharp distinction between energetical and cognitive factors (Hull 1943). According to the Yerkes–Dodson law, cognitive processing and its energetical supply are only loosely related through different optima of arousal for different levels of task complexity. Again, stress reactions were supposed to be essentially of the same nature, irrespective of a state of over- or under-arousal. The unidimensional view has been challenged with regard to arousal as well as to stress. It has been argued, both from physiological (Lacey 1967, Näätänen 1973) and behavioural (Hamilton et al. 1977, Hockey 1979) evidence that the organism should be conceived of as being in one of several activation states, which are intimately related to ongoing cognitive operations. With regard to stress one might be reminded of Appley and Trumbull's (1967) question whether stress has done anything more than replacing concepts like anxiety, conflict and frustration.

However valid this criticism of the unidimensional view, it should still be noted that a description of energetical phenomena in terms of qualitatively different arousal patterns, undergoing qualitative changes under sub-optimal conditions through strategical shifts in resource allocation (e.g. Rabbitt 1979), does not easily lead to a testable theory of arousal and stress. Of course it could be that such an undertaking is premature and that the aim should be limited to mapping out possible energetical–cognitive patterns. Yet this paper will attempt to follow an alternative route characterized by linear stage models. It will be argued that this may lead to perhaps a limited but a more testable view on energetical–cognitive patterns, as well as to a clearer view on how they could relate to stress. The proposed view consists basically of combining some of the results emerging from research on processing stages in choice reactions (Sternberg 1969, Sanders 1980) with Pribram and McGuinness' (1975) energetical concepts together with some additional elements which are on Kahneman's (1973) credit.

3. Outline of the view

Following the additive factor logic, a reaction process is conceived of as being subdivided into a number of serial processing stages, each of which has a specific contribution to the translation from a signal into a response. The various assumptions and pros and cons of this approach are discussed elsewhere (Sanders 1980). Here it may suffice to say that in principle the operations contained in each stage are supposed to be structural but, in addition, the efficiency of processing is affected by the state of the subject, albeit probably to a varying degree and in a different way for different stages. On the basis of the evidence, Sanders (1980) distinguished six stages in the traditional choice–reaction process, but for convenience the argument will be limited to those stages that seem best established *in casu*; stimulus preprocessing (signal intensity), feature extraction (signal quality) response choice (S-R compatibility) and motor adjustment (time uncertainty).

Effects of the energetical state of the subject are supposed to only appear to the extent and in the way *active* processes play a role in the cognitive operations of each stage. When considering the typical experimental variables affecting processing duration at the various stages, it seems likely that different types of active process are involved. Thus, with regard to time uncertainty the efficiency of processing is affected by preparatory processes and timing, which can be assumed to preset the involved stage (motor adjustment) as close as possible to the 'motor action limit' (Näätänen and Merisalo 1977) at a given moment in time. In contrast, with regard to S–R compatibility, the involved activity is concerned with adequate handling of the decision rules as imposed by instruction, which comes close to reasoning. With regard to signal quality, active processes may have the primary role of separating irrelevant and relevant elements of the percept. This may be especially the case when a passive 'global' analysis does not suffice to produce a complete output of the feature extraction stage (Navon 1977, Broadbent and Broadbent 1980). Finally, stimulus intensity may only give rise to 'bottom-up' processes, which is not to say that these are irrelevant for the energetical state. On the contrary, the collative variables (Berlyne 1969) have been traditionally considered as enhancing arousal or at least evoking a phasic response.

Apart from these differences in type of activity, it is probably of interest that, the later a stage in the sequence, the more its active processes occur primarily *prior* to the presentation of the stimulus. This is very clear for presetting in preparatory processes, but it also occurs in response choice in particular in cases of S–R frequency imbalance. Thus, it is not tenable that, as Rabbitt (1979) has recently argued, linear stage models are 'data-driven' by their very nature. They certainly allow for 'resource-driven' processes, although the stage structure puts constraints which are absent in free allocation resource models. The types of constraint may be quite in line with those met in multiple capacity theory (Sanders 1979, Navon and Gopher 1979), the active processes requiring sufficient energetical resources in order to enable adequate functioning, but each needing those that fit the nature of that specific type of active process.

What kind of energetical processes, then, might be considered? Following the earlier discussion, at least three would qualify. One would be a rather stimulus-independent type of resource, enabling anticipatory activity and pre-setting and related to readiness to respond. A second type of resource would be related to the active analysis of stimuli, and would therefore be much more bound to their actual presence. Finally, resources should be related to the control of response choice and handling decision rules.

There are striking resemblances between these deductions and the Pribram and McGuinness theory about the control of attention. On the basis of extensive neurophysiological evidence, they assume three systems, arousal as a phasic response to input, activation as a tonic readiness to respond, and effort as a co-ordinating principle between arousal and activation. Whenever the state of arousal and/or activation is sub-optimal—either too high or too low—the effort system is capable of adjusting the parameters, at least to some extent and for some time. At too high a state of arousal or activation, effort aims at preventing inadequate automatic stimulus–response sequences which could occur through the connection between arousal and activation, bypassing the cognitive processing stages. This is prevented by uncoupling arousal and activation as well as by attempts to moderate their state. When the states are too low, effort plays a stimulating role to promote activity which might otherwise not occur. Hence, the aim is to guarantee well considered behaviour and thus to increase the competence of the information-processing system. Pribram and McGuinness explicitly relate effort to 'reasoning' so that, next to co-ordinating arousal and activation, effort might directly provide resources to the response-choice system. Thus, it is a central-control system of behaviour, while arousal and activation are in principle more endogenous, sensitive to habituation and to general 'state' factors like diurnal rhythm.

In order to be informed about the state of arousal and activation there must be some evaluation mechanism as a feedback for the appropriateness of coping. In principle, two types of feedback might be available: one might directly reflect the physiological state of the system, so as to guarantee immediate action of the effort mechanism in case of unacceptable imbalance. Another one might reflect performance, or more generally, the extent of coping on a cognitive level. The evaluation system must dispose of criteria for comparison with the actual state. The physiological criteria depend on the ranges of variation which the bodily mechanisms can bear. On the cognitive level, the criteria are determined by the demands imposed by the environment and by personal goals. In other words, the cognitive criteria contain important motivational determinants. *Stress arises in this framework* (figure 1) *when the effort mechanism is either seriously loaded or falls short in providing the necessary adjustments.* It is clear from figure 1 that various patterns may arise, depending on where the problems are predominant. For example, Welford (1973) discussed three major types of deviation from optimum conditions, one relating to the amount of stimulation (arousal), another one to the predictability of events (activation) and a final one to conflict of data, promoting uncertainty in decision-making (effort). These deviations can be in either direction and may therefore represent six different patterns of stress in conditions of actual ongoing performances.

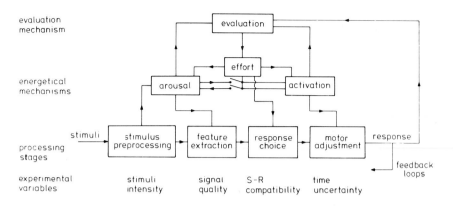

Figure 1.

4. Consequences of the working model

A first major consequence of this argument is that stress is always due to an attempt to cope. This is in line with Selye's (1956) classical stress hormone, corticosteroid excretion, acting as an energy mobilizer. At the same time, it has the implication that performance decrement is not a satisfactory index of stress. On the contrary, a decrement of performance may mean nothing but a failure of the criteria to note the divergence, or a decision that it is not relevant to cope. This may apply in particular to monotonous situations, where arousal and activation are presumably low. It leads to the conclusion that subjects who show least performance decrement—and thus attempt to cope—are most under stress (O'Hanlon 1981). Space prohibits discussion of the pertinent literature on this issue. Here it should suffice to note the relation with both personality traits and with environment-related measures like pacing speed and knowledge of results, which should naturally stimulate performance as well as increase stress (e.g. Frankenhaeuser 1977, Johansson *et al.* 1978, Stanger 1977).

The effects of too high a state of arousal and activation may be much more directly related to performance decrement. Moderation of the states of arousal and activation might take away resources for adequate decision-making. This is exactly what intuitively happens in a nervous state. It is an under-researched area, probably due to the fact that creating the pertinent conditions easily meets ethical constraints. Yet the work of Ursin *et al.* (1978) on student paratroopers shows that research into this direction is in principle possible.

A second consequence of the model of figure 1 regards effects of sub-optimal conditions which are predicted to be selective rather than general. Following the idea that energetical supply and cognitive processing are strongly related, the implication is that sub-optimal conditions have specific points of application on either arousal, activation or effort, showing themselves in decrements of either perception, action or response choice. There are now several results supporting this suggestion. Thus, the effect of amphetamine on choice–reaction processes seems largely limited to preparatory processes and to movement execution. Amphetamine also counteracted the usual negative effect of time-on-task on performance, and more so after sleep loss. Sleep state had also strong interactions with time uncertainty, and with time-on-task, both of which are fully removed by amphetamine. Both sleep loss and amphetamine showed hardly any relation with variations in S–R compatibility, suggesting that response choice is not affected, but the state of activation and its accompanying motor processes are. On the other hand, the effects of barbiturates were not found to be related to variations in either time uncertainty or S–R compatibility, but interacted with the effect of signal quality, suggesting that the arousal mechanism is selectively affected by barbiturates.

These data are largely based on recent work by Frowein (1981a, 1981b). In addition, Sanders *et al.* (1981) found that the effects of sleep state and signal quality interact, from which it is implied that lack of sleep depresses arousal as well as activation. All these data clearly indicate selective effects of suboptimal conditions on processing information and a main task for future research is to study the extent to which stable patterns of similar energetical—cognitive relations show up. Obviously this research should not be limited to behavioural variables, but should also include cortical, autonomous and hormonal variables.

As argued earlier, these patterns do not yet imply stress which is only supposed to occur when adverse effects are counteracted. In the above-discussed studies, it is unknown to what extent that was the case. Yet there are usually casual observations that some subjects show surprisingly little effects of lack of sleep, which is subsequently ascribed to motivation. Again, Wilkinson (1969) has shown that proper incentives like knowledge of results, which keeps the evaluation mechanism constantly aware of the quality of performance, have the effect of removing all effects of lack of sleep. Hence there are experimental tools available to vary the activity of the effort mechanism. Apart from knowledge of results one might think of varying personality type, for example the A/B typology, which is now often related to the likelihood of cardiovascular diseases—pacing speed, social pressure, or monetary rewards. It is a highly intriguing possibility that the type of stress response depends on the specific task for the effort mechanism. Is there a different hormonal pattern when the arousal mechanism or activation mechanism is either enhanced or moderated? Does this relate in any way to the type of illness following long-term exposure to stress? Future research will decide, but at least the working model suggests predictions for various specific performance studies.

References

APPLEY, M. H., and TRUMBULL, R., 1967, On the concept of psychological stress. In *Psychological Stress: Issues in Research,* edited by M. H. Appley and R. Trumbull (New York: Appleton).

BERLYNE, D. E., 1969, The development of the concept of attention in psychology. In *Attention in Neurophysiology,* edited by C. R. Evans and T. B. Mulholland (New York: Appleton).

BROADBENT, D. E., 1971, *Decisions and Stress* (London: Academic Press).

BROADBENT, D. E., and BROADBENT, M. H. P., 1980, Priming and the passive/active model of work recognition. In *Attention and Performance,* edited by R. S. Nickerson (Hillsdale N. J.: Erlbaum).

FRANKENHAEUSER, M., 1977, Job demands, health and wellbeing. *Journal of Psychosomatic Research,* **21,** 313-321.

FROWEIN, H. W., 1981a, Selective effects of barbiturate and amphetamine on choice reaction and movement time. *Acta Psychologica* (in press).

FROWEIN, H. W., 1981b, Effects of two counteracting stresses on the reaction process. In *Attention and Performance,* edited by A. D. Baddeley and J. Long (in press).

HAMILTON, P., HOCKEY, G. R. J., and REJMAN, 1977, The place of the concept of activation in human information theory; An integrative approach. In *Attention and Performance,* edited by S. Dornic (Hillsdale N.J.: Erlbaum).

HEBB, D. O., 1955, Drives and the CNS (conceptual nervous system). *Psychological Review,* **62,** 243-254.

HOCKEY, G. R. J., 1979, Stress and the cognitive components of skilled performance. In *Human Stress and Cognition,* edited by V. Hamilton and D. M. Warburton, pp. 141-171.

HULL, C. L., 1943, *Principles of Behavior* (New York: Appleton Century).

JOHANSSON, G., ARONSSON, G., and LINDSTRÖM, B. O., 1978, Social psychological and neuroendocrine stress reactions in highly mechanised work. *Ergonomics,* **21,** 583-600.

KAHNEMAN, D. A., 1973, *Attention and Effort* (New York: Prentice Hall).

LACEY, J. I., 1967, Somatic response patterning and stress: Some revisions of activation theory. In *Psychological Stress: Some Issues in Research,* edited by M. H. Appley and R. Trumbull (New York: Appleton).

LAZARUS, R. S., 1967, *Patterns of adjustment* (New York: McGraw-Hill).

LEVI, L., 1971, Stress, distress and psychosocial stimuli. In *Occupational Stress,* edited by A. McLean (Springfield, USA: Thomas).

NÄÄTÄNEN, R., 1973, The inverted-U relationship between activation and performance: A critical review. In *Attention and Performance,* edited by S. Kornblum (New York: Academic Press).

NÄÄTÄNEN, R., and MERISALO, A., 1977, Expectancy and preparation in simple reaction time. In *Attention and Performance,* edited by S. Dornic (Hillsdale, N.J.: Erlbaum).

NAVON, D., 1977, Forests before trees: The precedence of global features in visual perception. *Cognitive Psychology,* **9,** 353-383.

NAVON, D., and GOPHER, D., 1979, On the economy of the human information processing system. *Psychological Review,* **86,** 214-255.

O'HANLON, J. F., 1981, Boredom: practical consequences and a theory. *Acta Psychologica* (in press).

PRIBRAM, K. H., and McGUINNESS, D., 1975, Arousal, activation and effort in the control of attention. *Psychological Review,* **82,** 116-149.

RABBITT, P. M. A., 1979, Current paradigms and models in human information processing. In *Human Stress and Cognition,* edited by V. Hamilton and D. M. Warburton.

SANDERS, A. F., 1979, Some remarks on mental load. In *Mental Workload,* edited by N. Moray (London: Plenum).

SANDERS, A. F., 1980, Stage analysis of reaction processes. In *Tutorials in Motor Behavior,* edited by G. Stelmach and T. Requin (Amsterdam: North Holland).

SANDERS, A. F., WIJNEN, J. L. C., and ARKEL, A. E., 1981, An additive factor analysis of the effects of sleep-loss on reaction processes. *Acta Psychologica* (in the press).

SELYE, H., 1956, *The stress of life* (New York: McGraw-Hill).

STANGER, R., 1977, Homeostasis, discrepancy, dissonance: a theory of motives and motivation. *Motivation and Emotion, 1,* 103–138.

STERNBERG, S., 1969, On the discovery of processing stages: Some extensions of Donders' method. In *Attention and Performance II, Acta Psychologica, 30,* 276–315.

URSIN, H., BAADE, E., and LEVINE, S., 1978, *Psychobiology of Stress: A Study of Coping Men* (New York: Academic Press).

WELFORD, A. T., 1973, Stress and performance. *Ergonomics, 16,* 567–580.

WILKINSON, R. T., 1969, Some factors influencing the effect of environmental stressors upon performance. *Psychological Bulletin, 72,* 260–272.

Section 3.   Variables relating to stress

Relationship of perceived stress to job satisfaction

By T. K. Sen,[†] S. Pruzansky[‡] and J.D. Carroll[‡]

[†]American Telephone and Telegraph Company,
Morristown, New Jersey 07960, USA.

[‡]Bell Laboratories, Murray Hill, New Jersey 07974, USA.

Some preliminary findings are presented of a study relating perceived work stress to both overall job satisfaction and some of its psychological dimensions. Job autonomy and job demand were found to be significantly correlated to overall stress and overall satisfaction.

1. Introduction

In recent years occupational stress has become one of the most frequently mentioned topics in the psychological and medical literature (Adams *et al.* 1980, Harrell 1980). Amongst other things, stress has been found to influence performance (Cohen 1980) and to be related to a number of workplace variables like task characteristics (Hackman and Lawler 1971, Bunker 1980), leader behaviours (House and Mitchell 1974), organizational climate (French and Caplan 1973), interpersonal relations (Schuler 1979) and physical environment (Manning 1965). This paper presents some preliminary findings of a study relating perceived work stress to both overall job satisfaction and some of its psychological dimensions.

2. Method

2.1. *Data source and subjects*

The data for this study came from an employee survey in the area of work relationships conducted in six USA companies. About 15 000 employees participated. However, the results presented here represent the response of about 5000 management employees. Further analysis is in progress with the remaining data.

The survey questionnaire covered the following general categories of work relationships: overall job satisfaction, work stress, physical environment and equipment, work methods, advancement, training, pay and benefits, supervisory and coworker relations, pace and impact of change, and employee perception of top management.

2.2. *Analysis*

A subset of 33 items in the questionnaire were selected for a multi-dimensional scaling analysis. These items showed significant correlation with one or both of two items dealing with overall perceived stress and job pressure. The items read as follows: (1) I feel pressured to achieve objectives ('get numbers') or meet deadlines I know I can't achieve. (2) I feel under more stress on the job now than I ever have before.

In performing the multidimensional scaling analysis, product moment correlation coefficients were used as measures of similarity. For each company the data was first subdivided for lower and higher management, then all the higher management data were put together into one group because of small numbers. Thus, the input to the multidimensional scaling analysis consisted of seven intercorrelation matrices, one for each of the six lower management groups and one for the higher management groups across all six companies.

A method called INDSCAL developed by Carroll and Chang (1970) was used for the multidimensional scaling analysis. This method was implemented by a computer program called SINDSCAL developed by Pruzansky (1975). In applying the INDSCAL model, it is assumed here that for each individual subgroup (six lower and one higher management), measures of similarity (in this case, correlations) are inversely related to distance between questionnaire items in a multidimensional judgement (or psychological) space. Each subgroup is assumed to have its own 'private' judgement space, but these separate 'private' spaces are not completely different from each other. There is commonality of dimensions as well as diversity in the pattern of what is known as 'relative saliences' for the dimensions of the judgement space. The relative salience of a dimension can be defined in terms of how much a change on that dimension affects the similarity judgement (here the correlation coefficients) of a subgroup.

In dealing with this notion of relative salience, the INDSCAL model assumes that each subgroup's private space is derived from a common space for the entire group. Mathematically, this is done by stretching (or shrinking) the dimensions of the group space in proportion to the subgroup's relative saliences, that is, the importance that each subgroup places on each dimension. For example, a stretching factor of zero will mean that the subgroup attaches no importance to that particular dimension. The subgroup may or may not do this consciously, that is, the attribute corresponding to the dimension may either not be perceived at all or the subgroup may act as if it is not perceived as important.

A five-dimensional solution, accounting for about 66% of the variance, produced four interpretable dimensions as follows: job autonomy, job demand, work practices and coworker relations. The fifth dimension was not clearly interpretable.

Next, a multiple regression analysis was performed between these dimensions as predictor variables and four separate criterion variables: two questionnaire items by two job categories. The two questionnaire items are: (1) I feel under more stress on the job now than I ever have before (referred to as overall perceived stress), and (2) All things considered, how satisfied or dissatisfied are you with your current job? (referred to as overall job satisfaction). The two job categories are: (1) lower management (LM) and (2) higher management (HM), across all six companies.

3. Results

Only two of the five dimensions, job autonomy and job demand, were found to be significantly correlated to overall stress and overall satisfaction (by significance it is meant that regression coefficients were statistically different from zero). Figure 1 displays the plane (from the five-dimensional solution) that corresponds to these two dimensions. The horizontal axis represents the job-autonomy dimension and the vertical axis the job-demand

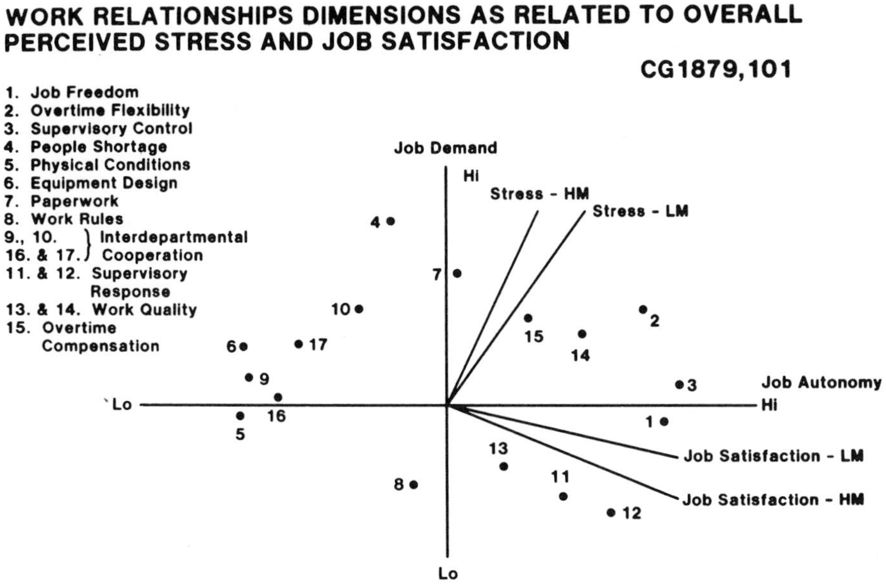

WORK RELATIONSHIPS DIMENSIONS AS RELATED TO OVERALL PERCEIVED STRESS AND JOB SATISFACTION

CG1879,101

Figure 1. INDSCAL item weights plane, showing two significant work relationships dimensions and their relationships to overall stress and overall job satisfaction for two groups of managers.

dimension. The dots represent the location of the items in this two-dimensional space (out of a five-dimensional solution). Only 15 of the 33 items are shown here because of the relationship to these five dimensions. The projection of these items on each of the two dimensions indicate the degree of their relationship to the corresponding dimension. For example, items 1 and 3 are strongly related to high job autonomy, but have little relationship to the job-demand dimension. Similarly, items 4 and 7 show a strong relationship to high job demand, but very little relationship to the job-autonomy dimension.

The four vectors represent the two item variables, overall stress and overall job satisfaction for each of the two managerial groups. The cosine of the angle between a vector and a dimension indicates the correlation (approximately) between those variables. Therefore, the smaller the angle, the higher is the correlation. For example, job satisfaction shows a little higher correlation with job autonomy than with job demand. Stress, on the other hand, shows more or less equal correlation with both dimensions.

Whereas figure 1 is a two-dimensional space for item weights, figure 2 is a similar weights space for the 'judges', which here correspond to the six subgroups described above. This figure shows how each subgroup ascribes importance to the two dimensions. For example, the lower management people in company 1 (L_1) place more importance on job demand or are more concerned about it than job autonomy. On the other hand, lower management people in company 5 (L_5) show more concern about job autonomy than job demand. Given limited resources, plots like this can help managers identify areas of concern of higher priority for action planning.

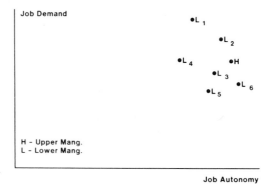

**WORK RELATIONSHIPS DIMENSIONS:
WEIGHTS SPACE FOR JUDGES (COS.)**

Figure 2. INDSCAL judges' weights space, showing how each subgroup of managers ascribes importance to the two dimensions—job demand and job autonomy.

Given the current interest in occupational stress in most corporations, one question that is frequently asked is how is stress related to job satisfaction. If we examine figure 1 again, we find that the stress and job-satisfaction vectors are almost orthogonal, that is, uncorrelated. While this is an INDSCAL model output, the actual product moment correlation of the two items (job satisfaction and stress) in the raw data (for all managers) was also very low, about 0·16. This however, is not a surprising finding (see Karasek 1979).

We were also interested in seeing if there were differences in response patterns between these two items as a function of job groups. In this case, we looked at responses from both management and non-management employees. Respondents were separated into three job categories, non-craft or clerical, crafts and management. An examination of contingency tables did reveal some differences as shown in figures 3, 4 and 5. The number of lines within each cell of the 3×3 matrix is proportional to the frequency in the corresponding cell.

CONTINGENCY TABLE

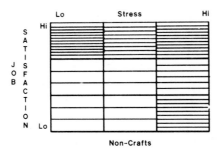

Figure 3. Contingency table, showing the response pattern of non-crafts or clerical people between two questionnaire items on overall stress and overall job satisfaction. The number of lines in each cell are proportional to the frequency in that cell.

CONTINGENCY TABLE

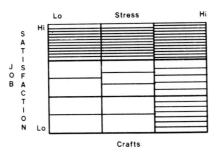

Figure 4. Contingency table, showing the response pattern of crafts people between two questionnaire items on overall stress and overall job satisfaction. The number of lines in each cell are proportional to the frequency in that cell.

T. Sen, S. Pruzansky and J. Carroll

CONTINGENCY TABLE

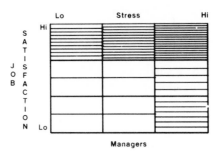

Managers

Figure 5. Contingency table, showing the response pattern of managers between two questionnaire items on overall stress and overall job satisfaction. The number of lines in each cell are proportional to the frequency in that cell.

Figures 3 and 4 show data for non-craft (or clerical) and crafts respectively, while figure 5 shows data for managers. All three groups show high job satisfaction and high stress; managers, however, show more association between high stress and high job satisfaction than the crafts and clerical people. Low stress is not generally associated with low satisfaction; on the other hand, low satisfaction is associated with high stress.

Further analysis is in progress aimed at better understanding of the relationship between perceived stress and job satisfaction, and at exploring the existence of other moderator variables, especially demographic variables like age and length of service. This may help, especially in predicting the level of job satisfaction within the high-stress group.

References

ADAMS, H. E., FEUERSTEIN, M., and FOWLER, J. L., 1980, Migraine headache: review of parameters, etiology, and intervention. *Psychological Bulletin*, **87**, 217–237.

BUNKER, K. A., 1980, *Perceived work stress and its effect on the emotional and physical well-being of Bell System Managers.* Basic Human Resources Research Report, AT&T, Morristown, N. J. 07960.

CARROLL, J. D., and CHANG, J. J., 1970, Analysis of individual differences in multi-dimensional scaling via an N-way generalization of Eckart-Young decomposition. *Psychometrika*, **35**, 283–319.

COHEN, S., 1980, Aftereffects of stress on human performance and social behavior: a review of research and theory. *Psychological Bulletin*, **88**, 82–108.

FRENCH, J. R. P., and CAPLAN, R. D., 1973. Organizational stress and individual strain. In *The Failure of Success*, edited by A. J. Murrow (New York: AMACOM), 30–66.

HACKMAN, J. R., and LAWLER, E. E., 1971, Employee reactions to job characteristics. *Journal of Applied Psychology Monograph*, **55**, 259–286.

HARRELL, J. P., 1980, Psychological factors and hypertension: a status report. *Psychological Bulletin*, **87**, 482-501.

HOUSE, R. J., and MITCHELL, T. R., 1974. Path-goal theory of leadership. *Journal of Contemporary Business, 3*, 81-97.

KARASEK, R. A., Jr, 1979, Job demands, job decision latitude, and mental strain: implications for job redesign. *Administrative Science Quarterly*, **24**, 285-308.

MANNING, P., 1965, *Office design: a study of environment* (Liverpool: University of Liverpool).

PRUZANSKY, S., 1975, *How to use SINDSCAL: a computer program for individual differences in multidimensional scaling*. Bell Laboratories, Murray Hill, N. J., 07974.

SCHULER, R. S., 1979, A role perception transactional process model for the communication-outcome relationships. *Organizational Behaviour and Human Performance*, **23**, 268-291.

Stress-related effects in the assessment of synthetic-work performance

By B. B. Morgan, Jr
Center for Applied Psychological Studies, Department of Psychology,
Old Dominion University, Norfolk, Virginia 23508, USA

The synthetic-work methodology has been employed for over 20 years to assess the effects of stresses such as work–rest schedules, sleep loss, continuous work and illness on complex operator performance. Conclusions drawn from this programme of research are summarized here in order to provide guidelines for future stress research. This summary indicates that future research should concentrate on the assessment of work-performance rather than test-performance responses to occupational stresses. It also suggests that additional research is needed to investigate the patterns of stress-related responses produced by different types of stresses of different severity, individual differences, and interacting variables such as performance strategies, the circadian rhythm, and extramural demands. The need for studies of the recovery from stress effects is also discussed.

1. Introduction

As previous authors have indicated (cf. McGrath 1970, Cox 1978), one of the major approaches to the study of stress treats stress as a dependent variable. Researchers who have adopted this 'response-based' approach have measured stress-related responses in terms of (1) human health and safety (e.g., Rogg 1961, Friedman and Rosenman 1974), (2) man's mental and subjective state (e.g., Driscoll 1970, Kagan 1975), (3) the body's biochemical and physiological responses (e.g., Frankenhaeuser et al. 1971, Froberg et al. 1972), and (4) worker performance and productivity (e.g., Lazarus et al. 1952, Morgan 1974, Alluisi and Fleishman 1981).

While each of these types of response is important in its own right, the last-mentioned takes on added significance because of the hypothesis that stress responses of the first three types are likely to be reflected ultimately in changes in performance and productivity. This chapter deals primarily with this latter type of response. Its purposes are to summarize a number of stress-related conclusions derived from previous investigations with the synthetic-work methodology, and to draw pertinent implications concerning requirements for future stress research.

2. The synthetic-work methodology

One of the most systematic programmes of research concerning the performance effects of stress has been based on the use of the synthetic-work approach to performance assessment (cf. Alluisi and Chiles 1967, Chiles *et al.* 1968, Alluisi 1969, Morgan and Alluisi 1972). The synthetic-work methodology has been employed for more than 20 years to assess the effects of stresses associated with work–rest cycles, the circadian rhythm, sleep loss, infectious diseases, continuous work, interrupted rest and recovery, noise, occupational exposure to inorganic lead, team-training loads, and the menstrual cycle on complex operator performance. To date, more than 30 major studies have been conducted, involving more than 39 000 subject-hours (nearly 19 subject-years) of data collection (a partial listing of these studies is presented in Alluisi *et al.* 1977).

The synthetic-work methodology provides for the measurement of sustained, time-shared performances in a laboratory-based synthetic job. It is designed to obtain measures of individual and team performances from five subjects who concurrently perform up to five of six tasks presented with a Multiple-Task Performance Battery (MTPB). These tasks were selected to measure performance functions required in a variety of jobs. Specifically, three watchkeeping tasks are used to measure the subjects' performances of watchkeeping, vigilance, and attentive functions (blinking-lights, warning-lights, and probability monitoring) and three active tasks are used to measure the performances of memory functions (arithmetic computations), sensory-perceptual functions (target indentification), and procedural functions (code-lock solving). Five of the tasks are individual-performance tasks, whereas one (code-locking solving) is a group-performance task.

They are performed according to a basic two-hour task programme which presents different combinations of three, four, or five tasks over successive 15-minute intervals. The task programme provides 30 minutes of low-demand performance, 60 minutes of intermediate-demand performance, and 30 minutes of high-demand performance during each two-hour period of testing. From the subject's viewpoint, there is no break between repetitions of the programme from the start to the end of a 'work day', because the three watchkeeping tasks are presented continuously at each workstation. In any given experiment, subjects are hired to work at this job just as they would at any other; they typically work eight hours per day for 12–15 days. Each of the MTPB tasks, as well as all other aspects of the synthetic-work methodology, have been described fully in previous reports (e.g., Chiles *et al.* 1968, Alluisi 1969, Morgan and Alluisi 1972).

3. Conclusions from past research

The following discussion presents only a few of the most important conclusions from previous investigations of the effects of work–rest schedules, sleep loss, continuous work and infectious diseases. The data on which these conclusions are based have been published previously (e.g., Chiles *et al.* 1968, Alluisi 1969, 1972, Alluisi *et al.* 1973, Beisel *et al.* 1974, Morgan 1974, Alluisi *et al.* 1977), and the reader is encouraged to review in detail the results presented in the cited publications.

3.1. *Work versus test-behaviour responses to stress*

When tested only briefly at specified intervals during or at the end of 48 hours of continuous activity, performance may be insensitive to the effects of sleep-loss stress (e.g., Ainsworth and Bishop 1971). On the other hand, when crews are required to work at the same job for 48 hours without sleep, large decrements are almost certain to occur (Morgan *et al.* 1974). These and other findings suggest that subjects approach work and performance-testing situations quite differently (Alluisi *et al.* 1977, Morgan 1980). Relative to the highly skilled performance of experienced workers, the performance required in most test situations is likely to be simple, unitary, brief, and essentially unskilled. A person's motivation and general approach to performance are likely to be quite different in these different situations; furthermore, the entire pattern of his reactions to stress are likely to be different in test and work situations. In order to obtain results that have the highest degree of relevance for operational situations, future stress research should be based as closely as possible on generalizable work-behaviour type measurement rather than on test-behaviour measurements.

3.2. *The pattern of responses to stress*

The performance effects of both continuous-work and infectious-disease stresses have been found to be general rather than specific to the type of performance measured. That is, these stresses affect all the individual-performance tasks of the MTPB in a similar fashion. These stresses produce general, systemic responses that are reflected in general physiological, psychological, and performance effects. On the other hand, some stresses (e.g., inorganic lead) have very specific effects that are only reflected in certain types of biochemical (e.g., aminolevulinic acid dehydrase) and performance measures (e.g., psychomotor performance) (Repko *et al.* 1975). Additional research is needed in order to determine the relative generality or specificity of the pattern of responses produced by different stresses (Alluisi 1975).

3.3. *Work-schedule demands*

For relatively short periods, motivated workers have the resilience to meet the demands of moderately stressful work–rest schedules (Adams and Chiles 1960, 1961, Alluisi *et al.* 1963). However, when crews are required to work continuously, decrements in work efficiency will average approximately 18%, 22%, and 34% during 36, 44, and 48 hours of continuous work and sleep loss, respectively (Morgan *et al.* 1974, 1977, Alluisi *et al.* 1977). Furthermore, the extent of these performance decrements has been found to correlate highly with subjective measures of anxiety, depression, hostility and subjective stress (Driscoll 1970). Combined with the finding that the performance, physiological, and subjective effects of illness are greater during the more severe rabbit fever than during the less severe sandfly fever (maximum performance decrements average approximately 30% during rabbit fever and 20% during sandfly fever) (Alluisi *et al.* 1973), these results indicate that the performance effects of acute stresses are directly related to the severity of the stress. Further research is needed to investigate the relationship between the severity of stress and the *pattern* of stress-related responses produced by the stress.

3.4. *Moderators of responses to stress*

Performance reserves. To the extent allowed by working conditions, workers will pace their performances so as to maintain performance reserves that will allow them to meet the demands of working conditions. Under normal conditions these reserves will be sufficient to meet expected performance demands. However, in following more demanding schedules, an individual might deplete his performance reserves and become less able to meet the demands of subsequent stressful conditions. Although this can result in rather catastrophic decrements in performance (Alluisi *et al.* 1964), it appears that different individuals may maintain greater reserves than others (Morgan *et al.* 1980). Further research is needed in order to identify the strategies employed in the management of performance reserves and to determine the interactive effects of these coping strategies on the stress-related responses of different individuals.

The circadian rhythm. The circadian rhythm, typically reflected in the cycling of physiological measures, may also be reflected in performance, particularly under conditions of high workload or low motivation. Even when workers are highly motivated, performance is likely to follow a circadian cycle during periods of stress (Alluisi *et al.* 1964, Morgan and Coates 1974). Feedback, motivation, activity schedules and other techniques may be used to reduce the circadian cycling of performance, but the maintenance of performance may require a considerably greater cost to the worker during the low portions of the circadian cycle; these greater costs may be reflected by changes in biochemical measures or other types of response. Future research should investigate the extent to which the circadian rhythm may interact with other stresses to determine the overall pattern of stress-related responses.

Extramural demands. Crews working only 8 hours/day, but who are required to meet the demands of everyday living in an urban community of moderate size, perform no better than crews working 12 hours/day in a confinement study involving no 'extracurricular' demands—i.e., no travelling to and from work, no shopping, no chores around the house, etc. (Alluisi 1969, 1972). Thus, it must be recognized that the demands of the social, family, and personal environments interact with the intended experimental demands to determine the pattern of responses to a given stressful situation. The *total demands* of both the work and leisure activities should be considered in the design of future stress studies, and the relative stress-related effects of various extramural demands should be investigated.

3.5. *Individual responses to stress*

Very large individual differences have been found in the performance effects of continuous-work and infectious-disease stress. In one group of eight subjects, the average performance effects of 36 to 44 hours of sleep loss ranged from no decrement to a decrement of nearly 30% below baseline (Morgan *et al.* 1980). Similarly, the performance effects of illness range from essentially no decrement in some subjects to an extreme of 20% decrement per °F rise in temperature with rabbit fever and about 14% decrement per °F of fever with sandfly fever (Alluisi *et al.* 1973). The existence of this very broad range of individual differences suggests the need for additional research to investigate both the range and consistency of individual differences in response to different types of stress. Additional information concerning the extent to which individuals respond the *same way* when exposed repeatedly to the same stress (or different stresses) is important not only to the understanding of stress-related effects, but also to the design of jobs and equipment and the development of stress-management paradigms. This would appear to be one of the most important areas of investigation for future stress research.

3.6. *Recovery from stress*

The amount of sleep required for the recovery of performance following periods of continuous-work stress interacts with the level of the initial stress, the duration of the rest-and-recovery period, and the circadian rhythm (Morgan 1974, Morgan and Coates 1974, Alluisi *et al.* 1977). These interactions point to the fact that stress-related recovery functions are at least as complex as are the initial effects of stress. While a considerable amount of research has been devoted to the study of stress effects, relatively little effort has been devoted to the measurement of human recovery functions (Harris and O'Hanlon 1972). Future stress research should place a strong emphasis on the investigation of recovery requirements. Such research is necessary in order to provide a data base for the development of viable programmes for the management of stress effects.

Acknowledgements

The author accepts full responsibility for the presentation of the material in this paper. However, he wishes to acknowledge his substantial indebtedness to the many individuals who have contributed to the programme of research from which this material is drawn. He particularly acknowledges the major contributions of his colleagues and coworkers Earl A. Alluisi and Glynn D. Coates.

References

ADAMS, O. S., and CHILES, W. D., 1960, *Human performance as a function of the work–rest cycle* (*WADD TR 60–248*) (Wright-Patterson Air Force Base, OH: Aerospace Medical Research Laboratory, Aeronautical Systems Division, USAF, NTIS No. AD-240-654).

ADAMS, O. S., and CHILES, W. D., 1961, *Human performance as a function of the work–rest ratio during prolonged confinement* (*ASD TR 61–720*) (Wright-Patterson Air Force Base, OH: Aerospace Medical Research Laboratory, Aeronautical Systems Division, USAF, NTIS No. AD-273-511).

AINSWORTH, L. L., and BISHOP, H. P., 1971, *The effects of a 48-hour period of sustained field activity on tank crew performance* (*TR No. 71-16*) (Arlington, VA, Human Resources Research Organization).

ALLUISI, E. A., 1969, Sustained performance. In *Principles of Skill Acquisition*, edited by E. A. and I. McD. Bilodeau (New York: Academic Press), pp. 59–101.

ALLUISI, E. A., 1972, Influence of work–rest scheduling and sleep loss on sustained performance. In *Aspects of Human Efficiency*, edited by W. P. Colquhoun (London: English Universities Press), pp. 199–216.

ALLUISI, E. A., 1975, Optimum uses of psychobiological, sensorimotor, and performance measurement strategies. *Human Factors*, **17**(4), 309–320.

ALLUISI, E. A., and CHILES, W. D., 1967, Sustained performance, work–rest scheduling, and diurnal rhythms in man. *Acta Psychologica*, **27**, 436–442.

ALLUISI, E. A., and FLEISHMAN, E. A. (editors), 1981, *Human Performance and Productivity*: *Stress and Performance Efficiency* (Hillsdale, NJ: Earlbaum Associates) (in the press).

ALLUISI, E. A., CHILES, W. D., HALL, T. J., and HAWKES, G. R., 1963, *Human group performance during confinement* (*AMRL TDR 63–87*) (Wright-Patterson Air Force Base, OH: Aerospace Medical Research Laboratory, Aeronautical Systems Division, USAF, NTIS No. AD-426 661).

ALLUISI, E. A., CHILES, W. D., and HALL, T. J., 1964, *Combined effects of sleep loss and demanding work–rest schedules on crew performance* (*AMRL TDR 64–63*) (Wright-Patterson Air Force Base, OH: Aerospace Medical Research Laboratory, Aeronautical Systems Division, USAF, NTIS No. AD-606 214).

ALLUISI, E. A., BEISEL, W. R., BARTELLONI, P. J., and COATES, G. D., 1973, Behavioral effects of tularemia and sandfly fever in man. *Journal of Infectious Diseases*, **128**, 710–717.

ALLUISI, E. A., COATES, G. D., and MORGAN, B. B. Jr, 1977, Effects of temporal stressors on vigilance and information processing. In *Vigilance*: *Theory, Operational Performance and Physiological Correlates*, edited by R. R. Mackie (New York: Plenum Press), pp. 361–421.

BEISEL, W. R., MORGAN, B. B. Jr, BARTELLONI, P. J., COATES, G. D., DERUBERTIS, F. R., and ALLUISI, E. A., 1974, Symptomatic therapy in viral illness: A controlled study of effects on work performance. *Journal of the American Medical Association*, **228**, 581–584.

CHILES, W. D., ALLUISI, E. A., and ADAMS, O. S., 1968, Work schedules and performance during confinement. *Human Factors*, 10(2), 143–196.

COX, T., 1978, *Stress* (Baltimore: University Park Press).

DRISCOLL, J. M., 1970, Psychological responses to continuous work under conditions of sleep loss. Appendix C in *Effects of 48 hours of continuous work and sleep loss on sustained performance (ITR-70-16)*, by B. B. Morgan, Jr., B. R. Brown, and E. A. Alluisi (Louisville, KY: University of Louisville, Performance Assessment Laboratory), pp. 67–81.

FRANKENHAEUSER, M., NORDHEDEN, B., MYRSTEN, A. L., and POST, B., 1971, Psychophysiological reactions to understimulation and overstimulation. *Acta Psychologica*, 35, 298–308.

FRIEDMAN, M., and ROSENMAN, R. H., 1974, *Type A Behavior and Your Heart* (New York: Knoft).

FROBERG, J., KARLSSON, C. G., LEVI, L., and LIDBERG, L., 1972, Circadian variations in performance, psychological ratings, catecholamine excretion, and urine flow during prolonged sleep deprivation. In *Aspects of Human Efficiency*, edited by W. P. Colquhoun (Cambridge: English Universities Press), pp. 247–260.

HARRIS, W., and O'HANLON, J. F., 1972, *A study of recovery functions in man (TM No. 10-72)* (Aberdeen Proving Ground, MD: U.S. Army Human Engineering Laboratory).

KAGAN, A., 1975, Epidemiology, disease, and emotion. In *Emotions: Their Parameters and Measurement*, edited by L. Levi (New York: Raven Press).

LAZARUS, R. S., DEESE, J. and OSLER, S. F., 1952, The effects of psychological stress upon performance. *Psychological Bulletin*, 49, 293–317.

McGRATH, J. E., 1970, *Social and Psychological Factors in Stress* (New York: Holt, Rinehart and Winston).

MORGAN, B. B., Jr, 1974, Effects of continuous work and sleep loss in the reduction and recovery of work efficiency. *American Industrial Hygiene Association Journal*, 35, 13–20.

MORGAN, B. B., Jr, 1980, *Influence of task and situational variables on the applicability of vigilance data to physical security*. Paper presented to the meeting of the Human Factors Society, Los Angeles, October.

MORGAN, B. B., Jr, and ALLUISI, E. A., 1972, Synthetic work: Methodology for assessment of human performance. *Perceptual and Motor Skills*, 35, 835–845.

MORGAN, B. B., Jr, and COATES, G. D., 1974, *Sustained performance and recovery during continuous operations (ITR-74-2)* (Norfolk, VA: Old Dominion University, Performance Assessment Laboratory).

MORGAN, B. B., Jr, BROWN, B. R., and ALLUISI, E. A., 1974, Effects on sustained performance of 48 hours of continuous work and sleep loss. *Human Factors*, 16(4), 406–414.

MORGAN, B. B., Jr, BROWN, B. R., COATES, G. D., and ALLUISI, E. A., 1977, Sustained performance during 36 hours of continuous work and sleep loss. *JSAS Catalogue of Selected Documents in Psychology*, 7, 91 (MS. No. 1560).

MORGAN, B. B., Jr, WINNE, P. S., and DUGAN, J., 1980, The range and consistency of individual differences in continuous work. *Human Factors*, 22(3), 331–340.

REPKO, J. D., MORGAN, B. B., Jr, and NICHOLSON, J., 1975, *Behavioral effects of occupational exposure to lead (Report No. (NIOSH) 75-164)* (Washington, D. C.: U.S. Department of Health, Education, and Welfare).

ROGG, S. G., 1961, The role of emotions in industrial accidents. *Archives of Environmental Health*, 3, 519–522.

The mechanism of mental fatigue

By MASAMITSU OSHIMA

Medical Information System
Development Center
Laddick Akasaka Bldg 2-3-4
Akasaka, Minato-Ku, Tokyo JAPAN

Extensive experimentation by the author suggests that critical flickering frequency is an effective methodology for measuring mental fatigue. A number of experimental studies are presented to illustrate this relationship.

We are now in the informatics age. Mental work is replacing muscular work as a result of mechanization and computerization. Because of this, we have to investigate mental work from the standpoint of health care, as mental fatigue is an indirect cause of health disturbances. In this paper, the results of my long-range investigation on the mechanism of mental fatigue will be presented. The critical flickering frequency (CFF) is available for the study of mental fatigue. The term 'flicker value' is also used in Japan instead of CFF. Of course, CFF and flicker value are the same. CFF means the tension level of human beings, and the tension level is the level of excitation. Human beings usually control the tension level for doing work.

CFF is in inverse proportion to the grade of sleepiness (table 1), and the grade of sleepiness is also in inverse proportion to the tension level, and so CFF has an intimate correlation with the tension level. Also, at the lower tension level, the error increases and productivity becomes lower.

Table 1. The relation between sleepiness and CFF

Subject	Coefficient of correlation	Significance of critical ratio (%)
1	-0.55	5
2	-0.76	5
3	-0.62	1
4	-0.29	10

86.0% declination of CFF corresponds to Sleepiness (7 Subjects)

The relation between CFF and workload is shown in figure 1. Workload corresponds to the ratio of usual daily amount of work to maximum daily amount of work estimated by short-timed maximum amount of work.

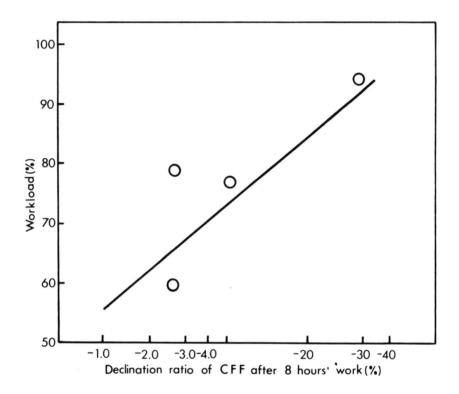

Figure 1. The relation between CFF and workload.

When someone begins to want to work he adapts the tension level to the required level to do work. When the work requires the higher tension level, the human being goes up the tension level to the necessary level. The initial going-up stage is shown in figure 2 but the tension level goes down before long. If the work is stopped at C-point, the tension level goes down as shown in the figure, and if the work is stopped at D-point, that goes down as shown in the figure. At D-point the tension level goes down further as compared with C-point. This fact shows that the latent difference between C- and D-point is clearly shown by the recovery situation (figure 2).

The recovery situation was investigated as one sample. Figure 3 shows the recovery situation at C-point and D-point. The recovery situation of relatively short driving time and relatively long driving time are shown in the figure. The former corresponds to recovery at C-point and the latter to recovery at D-point. The declination ratio of CFF is different in these two cases.

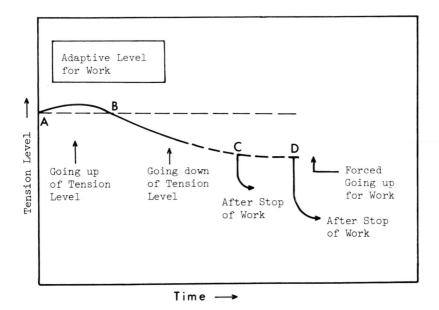

Figure 2. Sequential variation of tension level.

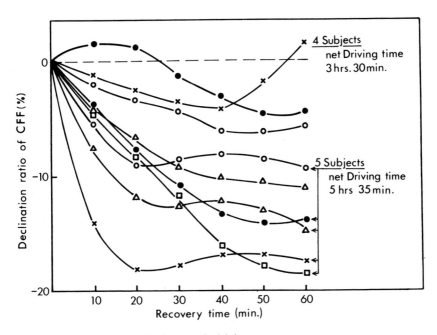

Figure 3. Recovery of CFF after truck driving.

M. Oshima

On the other hand, the recovery of CFF after a relatively short time of driving a motorbike is shown in figure 4.

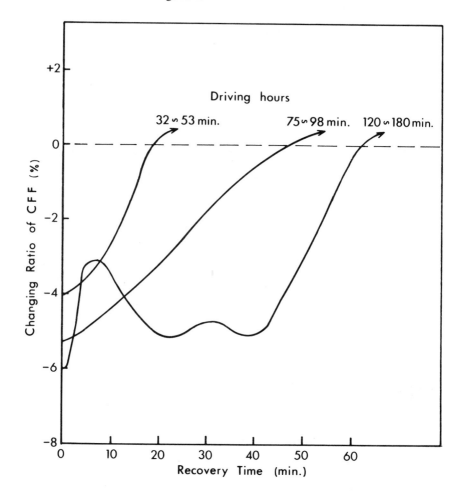

Figure 4. Recovery after riding a bicycle.

Next, the sequential variation of CFF until the next day is described. A human being has the 24 hours physiological rhythm. Figure 5 shows this rhythm, which is higher in daytime and lower at night. The higher phase in daytime is the stage of activity, and the lower phase at night is the stage of rest. The difference in reaction for the load of the bicycle-ergonometer is shown; that is to say, the action in the higher phase is less than that in the lower phase (figure 6). The over-reaction is shown in the lower phase, which exists at night.

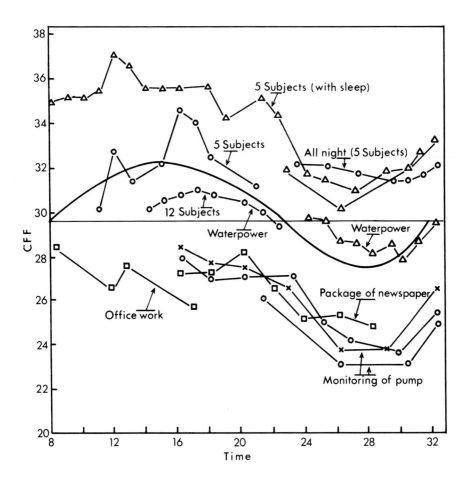

Figure 5. The situation of CFF in one day.

The sequential variation of CFF is shown in figure 7. The first is CFF before work, the second is that after the work of forenoon, the third is that before the work of afternoon, and the fourth is that after the work of afternoon. The difference between 1 and 1′ is estimated as the fatigue carried over to the next day (figure 7).

The variation of 24 hours physiological rhythm of mental fatigue is characterized by four items. The first is declination of baseline, the second is smaller fluctuation, the third is prolongation of one period, and the fourth is deviation of phase. In figure 8 the mechanism of the fatigue carried over to the next day can be observed by the declination of baseline and deviation of phase.

The daytime means the phase of activity and the stage for alternation of

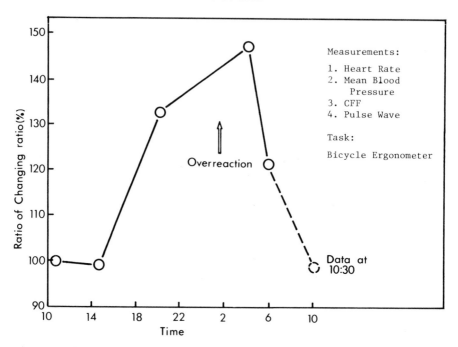

Figure 6. Changing ratio of some functions in each time.

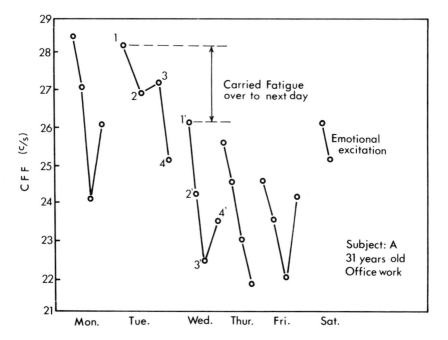

Figure 7. The sequential variation of CFF.

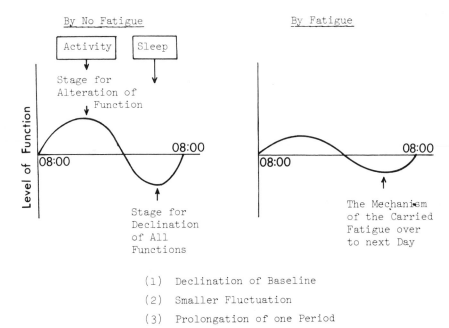

(1) Declination of Baseline
(2) Smaller Fluctuation
(3) Prolongation of one Period
(4) Deviation of Phase

Figure 8. 24 hours physiological rhythm.

Table 2. The causes of CFF declination being more than 10%

	Causes of CFF Declination	Cases
Prolongation of worktime	P. of semirestriction time	6
	Long-range exercise before work	5
	Shortening of unrestrained time	4
	Time factors	15
Unsatisfied recovery time	Shortening of sleeping time	9
	Long-range work at unrestrained w.	3
	Recovery time factors	12
	Disease	1

M. Oshima

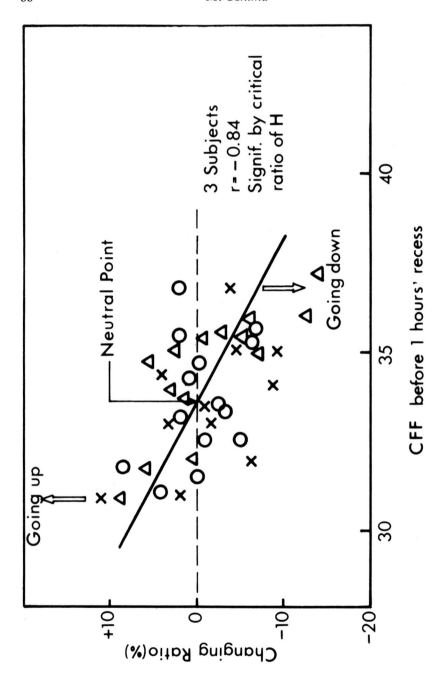

Figure 9. Change of CFF after one hour of recess.

function, by the mechanism of homeostasis, as shown in figure 9. CFF before one hours recess changes to maintain a level (neutral point) by an increase of the lower CFF and a decrease of higher CFF. This phenomenon shows the alternation of function related to antagonism at work and recess. This is one of the mechanism of homeostasis.

The causes of CFF declination, that is to say, mental fatigue are shown in table 2. In ordinary work the causes of mental fatigue are mainly the causes of time factor in almost all cases. This suggests that the time factor is more important than density of work.

References

OSHIMA, M. 1980. *On the Study of Fatigue*, Dobun-Shoin.

Attentive behaviour after exposure to continuous industrial noise

By K. LINDSTRÖM and S. MÄNTYSALO

Institute of Occupational Health, Helsinki, Finland

Attentive behaviour in a choice reaction-time task was studied in 11 workers exposed to continuous industrial noise and nine workers not exposed to such noise. The measurements were taken in three experimental sessions (before, in the middle, and after the workshift) in order to evaluate the possible aftereffects of noise and other workloads. Acute and chronic symptoms of strain were also investigated.

The group exposed to noise revealed a greater variability in reaction time than the controls. The slow reaction times of the exposed group were related to the acute strain ratings, while those of the control group were more dependent on chronic fatigue symptoms. The work of the exposed group also included more time pressure.

1. Introduction

Noise is one of the most common occupational exposures. Its extra-auditory adverse effects are not as well known as the related auditory health hazards; in particular, the psychological aftereffects of occupational exposure to noise are less known (e.g., McLean and Tarnopolsky 1977). It has generally been thought that the effects of noise on behaviour can best be seen from the aftereffects (Glass and Singer 1973), which are, however, predominantly dependent on other environmental factors and also on the nature of the work done (Poulton 1979).

Our purpose was to study the aftereffects of exposure to continuous industrial noise on attentive behaviour during the workday. Simultaneously, subjective evaluations of fatigue and other strain sensations were recorded. Special attention was paid to the relationship between acute and chronic symptoms of strain and the prolongation of reaction times.

We also investigated the perceived level of noise and collected other subjective evaluations on work. The relationship between these data and reaction times was assessed.

This chapter presents only some preliminary results of an extensive study.

2. Methods

2.1. *Subjects*

The group of subjects exposed to continuous noise comprised 11 male cable-factory workers, mainly machine operators, ranging in age from 22 to 39 years (mean ± S.D. $28 \cdot 2 \pm 4 \cdot 8$ years). Their tasks consisted of the inspection of both the quality of cable produced and the proper operation of various machines in a large factory hall, approximately 200×200 metres, with a high ceiling. Because the hall contained many large machines, the level of noise remained constant at 85–95 dB(A). The noise was steady and continuous with no noticeable intermittent bursts. Because the workers usually wore ear protectors, the individual exposure level remained lower than expected on the basis of the environmental noise level. Each work phase lasted from 1–3 hours, after which the operators had to change a reel in order to begin another similar work phase. The workers' duration of exposure to noise ranged from 2–10 years (mean ± S.D. $5 \cdot 9 \pm 2 \cdot 2$ years).

Nine healthy young men drafting the final plans for ships in the office of a shipyard served as the nonexposed control group. Their ages ranged from 17–28 years, with a mean of $23 \cdot 8$ years (S.D. $3 \cdot 6$).

2.2. *Measurement methods*

The attentive behaviour of the subjects was measured in a vigilance task with a visual choice-reaction time (RT). The RT measurements were recorded on two separate days just prior to the workshift in the morning, at noon, and at the end of the workday. In every session, a series of 225 stimuli were given. On the board in front of the subjects, three red lights (ø 3 cm) were lit in random order. The subjects had to switch off the light on the right or left, but not the one in the middle, as quickly as possible. The interstimulus interval varied randomly between two, four and six seconds.

Before each session began, the subjects subjectively rated their strain on a scale with the following five pairs of adjectives: relaxed–tense; calm–anxious; comfortable–uncomfortable; alert–fatigued; irritable–cheerful. They were also asked to complete a questionnaire which dealt with their past histories of work and health and their chronic symptoms.

2.3. *Statistical methods*

In the statistical analysis, the data from the first measurement day were used. From each session the first 25 RTs were excluded and the last 200 RTs included. The mean RTs and standard deviations of the groups were counted for each session and for the blocks of each session, each block containing 50 RTs. The inter- and intra-group differences were tested with the Student's t test. The differences in the dispersions were tested with Barlett's test for equality of standard deviations. The interactions and relationships between the RTs and other variables were analysed with the chi-square test or Pearson's r.

3. Results

3.1. *Reaction times*

Table 1 shows the mean RTs and standard deviations of the two groups. In each session, the mean RTs of the exposed group were longer than those of the control group, but the differences were not quite statistically significant. However, the RT variance of the exposed group was greater than that of the control group in the noon session ($p \langle 0 \cdot 10$).

Table 1. Mean reaction times and standard deviations of the groups.

Session	Exposed group ($N = 11$) Mean	S.D.	Control group ($N = 9$) Mean	S.D.
Morning	417	59	402	40
Noon	415	53*	390	33*
Afternoon	422	54	392	45
All sessions	418	53	395	37

*Differences significant at the 10% level (Barlett's test for equality of standard deviations).

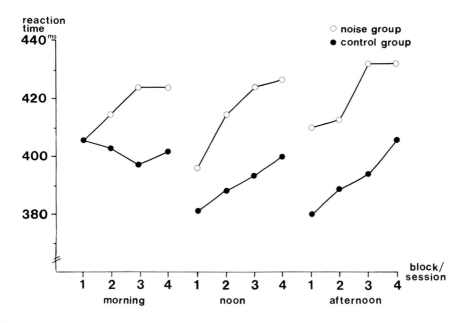

Figure 1. Mean reaction times of the groups in the three sessions divided into four blocks.

Figure 1 shows the mean RTs of the groups. During the course of a session, some trends could be seen. The speed of the reactions in the first quarter of the morning session of both groups was about the same. In the later quarters of the morning session the RTs remained nearly constant for the controls, while those of the exposed group became slower. In the noon and afternoon sessions the RTs of the exposed group were already slower at the beginning of each session; thereafter, the RTs of both groups slowed towards the end of the sessions. The differences were not statistically significant.

3.2. *Acute strain ratings*

Figure 2 shows the changes in the total amount of acute strain evaluated during the workday. The ratings were more positive at noon than in the morning or afternoon. In general, the strain ratings of the controls were more negative than those of the exposed group. In the control group, no relationship was found between the total amount of acute strain and the RTs, nor was the acute fatigue rating related to RT prolongation. However, in the exposed group the total amount of change in acute strain within the day correlated with corresponding change in the RTs ($r = 0.39$, $p < 0.05$, df $= 31$).

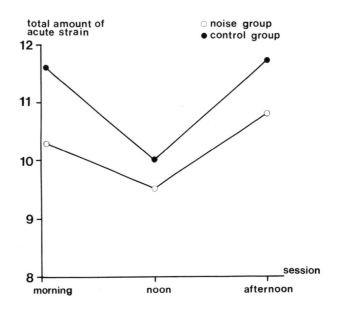

Figure 2. Mean of the sum of the acute strain ratings.

3.3. *Chronic symptoms*

The groups revealed no differences in three chronic symptoms—concentration difficulties, fatigue and anxiety. When these symptoms were related to the RTs, the control group showed a positive relationship between chronic fatigue and long RTs ($r = 0.47$, $p \langle 0.10$) (figure 3). This connection was not found in the exposed group. Conversely, concentration difficulties were related to long RTs in this group ($r = 0.52-0.62$, $p \langle 0.05$).

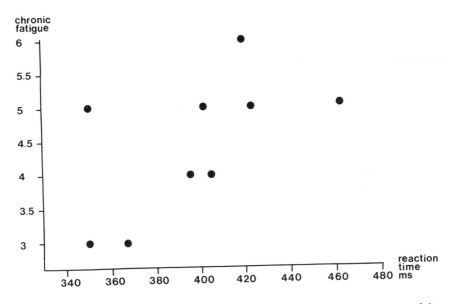

Figure 3. Relationship between the mean RTs and the chronic fatigue symptoms of the control group ($r = 0.57$, $p \langle 0.10$).

3.4. *Perceived work environment and work content*

The perceived level of noise was linearly related to the duration of exposure to noise ($r = 0.54$, $p \langle 0.10$) and time pressure experienced at work ($r = 0.64$, $p \langle 0.05$).

When compared to the control group, the exposed group evaluated their work as demanding less accuracy ($t = 2.25$, $p \langle 0.05$) but containing more time pressure ($t = 1.80$, $p \langle 0.10$). The two groups evaluated the mental load of their work similarly. The perceived mental workload was related to slow RTs in the exposed group; their relationship was the most clearly visible for the RTs measured in the afternoon ($r = 0.56$, $p \langle 0.10$). The slow RTs in the afternoon tended also to relate to perceived time pressure at work.

4. Discussion

The RTs of the exposed group differed from those of the control group. The former almost always had longer RTs than the latter, except in the beginning of the morning session before work. Their RT variances were also greater. In the exposed group, the RTs seemed to be prolonged already during the morning session, whereas in the control group the prolongation was seen later, especially in the afternoon session. Since the measurements were taken just after a work period, the results may primarily reflect the possible aftereffects of the total workload on attentive behaviour.

The acute sensations of strain, but also chronic concentration difficulties, can explain the slow reaction speed of the exposed group. These findings may be due to the effects of fatigue ensuing from the workload of the exposed group. In the control group the slow RTs were more dependent on chronic fatigue symptoms.

It has been proposed that a sudden decrement in performance occurs immediately after a noise load is removed. After noise exposure, arousal may first drop below normal and later recover (Poulton 1979). This kind of effect was not seen directly in our study. The RT task itself probably had a transient effect that increased the arousal of the group exposed to noise, because the workers changed from a monotonous inspection task to an RT task which demanded both high speed reaction and attention. Merely changing between two similar conditions can stimulate and increase arousal temporarily. However, because of the noise-induced strain and other loads related to the work, the activating effects of the change in conditions soon decreased, and the reaction speed slowed as the sessions proceeded.

Because the RT measurements were taken after the noise exposure and the groups were small, it is difficult to determine whether these findings were due to noise, monotony, time pressure at work, or to some combination of the three. In the group exposed to noise the workers could not control the noise exposure, and their tasks were moderately machine-paced, with time pressure. The controls had much more freedom in organizing their tasks, and their work environment was peaceful.

References

GLASS, D., and SINGER, J., 1973, Behavioural effects and aftereffects of noise. *Proceedings of the International Congress on Noise as a Public Health Problem, Dubrovnik* (Washington: U.S. Environmental Protection Agency Publication), 550/973-008, 587–592.

McLEAN, E. K., and TARNOPOLSKY, A., 1977, Noise, discomfort and mental health. A review of the socio-medical implications of disturbance by noise. *Psychological Medicine*, 7, 19–62.

POULTON, E. C., 1979, Composite model for human performance in continuous noise. *Psychological Review*, 86, 361–375.

Depression in the course of physical illness

By H. C. HENDRIE

Department of Psychiatry, Indiana University,
School of Medicine, 1100 West Michigan Street,
Indianapolis, Indiana 46223, USA

The prevalence of depression in physical illness is high. This chapter reviews some aspects of the relationships between physical illness and depression, and discusses the possible key role depression plays in understanding psychosomatic and somatopsychic mechanisms.

1. Introduction

The principles involved in the analysis of stress response are similar regardless of the type of stress studied. As physicians, we are particularly concerned about illness-generated stress. Kagan and Levi (1971) have constructed an elegant theoretical model for understanding psychosocially mediated illness (figure 1). Lipowski (1977) has emphasized the importance of emotional state as an intervening variable between the meaning of information to an

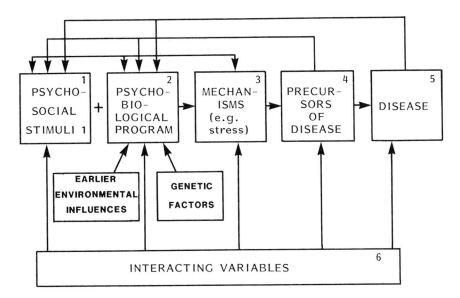

Figure 1. A theoretical model for psychosocially mediated disease (taken from Kagan and Levi 1971).

individual and elicited physiological change. Emotional state in this context could be considered as the equivalent of 'mechanisms' in the Kagan and Levi model.

At the Institute of Psychiatric Research, we are embarking on a combined basic and clinical programme to study psychosomatic and somatopsychic mechanisms. We are particularly interested in studying the apparently key role depression plays in these relationships. This chapter comprises a selective review of prior studies in this area.

2. Can depression cause physical illness?

The results of the existing studies are difficult to compare, mainly owing to the widely varying methodologies and differing definitions of depressive disorders utilized in them. Generally, retrospective studies involving already physically-ill patients have yielded more positive results than prospective studies which start with depressed or potentially depressed patient populations.

In a series of retrospective studies, Engel and his colleagues concluded that a significant proportion of patients with a variety of illnesses developed feelings of helplessness and/or hopelessness as a response to life stress experienced prior to illness onset (Schmale 1958, Engel and Schmale 1967). Engel (1968) later elaborated the giving up–given up complex which he felt could be crucial in illness development. He was careful to emphasize, however, that this complex by itself would not produce illness. It would do so only if the predisposition to illness already existed or potential pathogens were present. Recently, Murphy and Brown (1980) in a community survey came to a remarkably similar conclusion. Women who responded to life stress by developing a mild depressive disorder were much more likely, shortly thereafter, to become physically ill.

In contrast, Tsuang and Woolson (1978) in a long-term (30–40 years) follow-up of psychiatric in-patients diagnosed as suffering from depression could find no excess of mortality in these patients compared to population norms when suicides and accidental deaths were removed from the analysis. Similarly, Goldberg *et al.* (1979) in a 12-month follow-up community survey could find no relationship between initial ratings of depression and subsequent ill-health. However, Wigdor and Morris (1977) did find an excess of morbidity in a 20-year follow-up of depressed in-patients at a V.A. hospital compared to a matched population of paranoid patients. Kerr *et al.* (1969) also found a significant increase in the subsequent diagnosis of cancer in depressed elderly male in-patients. However, the time course was such that the authors concluded it was most likely that the depression was an early direct manifestation of cancer.

Prospective studies of bereavement sufferers yield similar, somewhat negative results. In an excellent review of such studies, Clayton (1979) concludes that only in the case of older widowers is there evidence for subsequent increased mortality. Although younger widows and widowers may report more physical distress and use more medication, there is little objective evidence of illness (Parkes 1970).

It is difficult to reconcile these results. With retrospective studies there is always the possibility that the apparently positive findings are the result of the human tendency in both patients and observers to try to find 'explanations' for illness. Some illnesses, e.g. cancer or hypertension, can exist for many years asymptomatically. Thus, as in the Kerr study, it is conceivable that the apparently pre-existing depression is rather a direct manifestation of the illness process. It is also possible that depression is simply too broad a concept to describe the potentially pathogenic psychic state. Some more specific formulation like that of Engel's may be more helpful. It is also conceivable that the depressed mood will only produce illness in the presence of other pathogens, e.g., lowering immunocompetence, which would in turn allow proliferation of hithertofore restrained neoplastic cells.

3. Can physical illness produce depression?

The number of illnesses reported to be associated with depression is staggering (for a review, see Hendrie 1978). The answer to this question would therefore appear to be obvious. However, a few words of warning are necessary.

The few carefully controlled prospective studies tend to be more cautious in their conclusions than the many retrospective and/or case studies. The identification of the depressive syndrome is particularly problematic in a physically-ill population. The study by Ganz *et al.* (1972) suggests that in the physically ill, symptoms referring to psychophysical states such as tiredness, insomnia, loss of energy, slowness, etc., are most likely due directly to the illness process and should be considered separately from symptoms such as depressed mood, guilt and hopelessness, which are more likely to reflect the depressive disorder. Community surveys of depression prevalence have also highlighted the necessity of distinguishing the common transient mood swings from the more persistent mood disturbances which are more indicative of a true depressive state (Craig and Van Natta 1979).

Despite these caveats, the vast preponderance of evidence supports the view that physical illness does act as an independent provocative agent in the genesis of depression (Roth and Kay 1956, Hendrie 1978).

4. What are the consequences
of being depressed for the physically ill?

It is surprisingly difficult to answer this question with any certainty. Mood fluctuates considerably over the course of an illness, and repeated measurements are thus necessary. However, few good longitudinal studies are available. There is also the problem previously mentioned in distinguishing between symptoms reflecting depression and symptoms directly due to illness; thus, apparent relationships between depression and illness can be spurious.

Most studies do tend to show a correlation between levels of depression and indices of illness activity. As the intensity of depression increases, so also does the severity of the illness (Shochet *et al.* 1969, Craig and Abeloff 1974). In Rimon's (1969) study of rheumatoid arthritics, patients who remained depressed over a six-month period tended to show also a deterioration in clinical status. Occasionally, studies report the opposite relationship, i.e., as somatic symptoms appear, depressive symptoms diminish. In a study of pain–mood patterns in rheumatoid patients, Molodofsky and Chester (1970) had an intriguing finding. About half of the patients demonstrated a 'synchronous' relationship between pain and mood, i.e., as pain intensity increased, so also did feelings of anger or anxiety. Half of the patients, however, demonstrated a 'paradoxical' relationship, i.e., as pain diminished, feelings of depression and hopelessness increased. In a 1–2 year follow-up, the paradoxical group had a less favourable clinical outcome.

There are a number of reports suggesting that therapy, either psychotherapy or pharmacotherapy aimed at alleviating depression, also leads to clinical improvement (Rimon 1974, Scott 1969, Shochet *et al.* 1969, Thorpe and Marchant-Williams 1974, Allen and Pitts 1978). Some authors suggest, however, that the most striking effects are on the subjective experience of pain or depression rather than on the objective manifestations of the disease process itself (Wright *et al.* 1963).

5. What are the mechanisms whereby depression
could exercise its effect on the illness process?

It is beyond the scope of this chapter to discuss in depth the potential mechanisms whereby depression could influence pathophysiology. Instead, let me mention briefly a few of the possibilities.

There has been a great surge of interest recently in the biology of depression. Many depressed patients show profound alterations in their sleep–waking cycle, not just involving more fragmented sleep patterns, but also affective sleep stages, particularly Stage IV sleep, the deepest stage, and Stage I R.E.M., or dreaming sleep (Chen 1979). A significant proportion of depressed patients, particularly those with the syndrome of melancholia, show evidence of hypersecretion in the adrenal-cortical system. It has been postulated that, unlike the case in the general stress response, this hypercortisolemia in depressed patients represents a fundamental disturbance of neurobiological` mechanisms involving the hypothalamus and other brain areas (Sachar *et al.* 1973, Carroll 1978). More recently, there has been considerable excitement generated by the possibility that stress and depression can alter immune mechanisms, e.g. bereavement has been associated with a depression in lymphocytic function (Bartrop *et al.* 1977, Rogers *et al.* 1979).

There has been surprisingly less interest in the potentially deleterious behavioural concomitants of depression. The motoric retardation often associated with depression could clearly negatively influence the outcome in disease affecting the joints or in the rehabilitation process following coronary artery disease. Feelings of helplessness, hopelessness, or a reduction in motivation could lead to non-compliance with treatment regimes, a major problem in medical care (Blackwell 1976).

6. Why do physically-ill patients get depressed?

The development of depression in response to illness would appear to be a very understandable reaction. Indeed, it could be argued with some justification that the question ought to be reversed. How do some seriously ill patients avoid becoming depressed? However, the relationship between them is more complex.

Table 1 lists some of the factors reported in the literature to be associated

Table 1. Factors associated with the production of depression in the physically-ill population.

Illness characteristics	*Environmental characteristics*
Types of illness	Concurrent life stress
Illness severity	Social-support system
Treatment environment	
Presence of 'organic' symptoms	
Pain	
Restriction of mobility	

Patient characteristics
Age
Sex
Previous personality
History of depression
Use of denial
Fear of dependency on others

with the production of depression in the physically ill. This table leaves out any consideration of the important medication effects. The factors are grouped together under three headings: illness characteristics, patient characteristics, and environmental characteristics.

While depression appears to be ubiquitous in all seriously ill patient populations, there does appear to be some evidence that certain illnesses are more associated with its production than others. Cancer, for instance, seems to be particularly depressogenic (Greer 1979, Whitlock and Siskind 1979). In some comparative studies, patients suffering from Systemic Lupus had a higher prevalence of depression than patients suffering from rheumatoid arthritis who in turn had a higher prevalence than osteoarthritics or patients with low-back pain, suggesting that systemic diseases are more likely to cause depression than localized disease processes (Ganz *et al.* 1972, Zaphriopoulos and Burry 1974). Severity of illness has been correlated with depression in some studies, but not in others (Moffic and Paykel 1975, Rogers *et al.* 1980). Surprisingly, most investigators find no relationship between length of illness and the incidence of depression. Pain intensity has most often been correlated with depression, but at least one author has found a negative relationship between pain and depressed mood (Molodofsky and Chester 1970, Moffic and Paykel 1975). Restriction of mobility has been associated with depression by several authors (Roth and Kay 1956, Moffic and Paykel 1975). The treatment environment has been mentioned as a possible provocative or preventive agent in the production of dysphoric moods in patients (Forester *et al.* 1978). This has obviously great implications for hospital construction and design. The concomitant presence in the patients of symptoms suggestive of brain dysfunction, i.e., mild confusion, memory dysfunction, etc., has also been associated with a high incidence of depression (Ganz *et al.* 1972).

There is some evidence, at least in cancer patients but also in gynaecological conditions, that younger patients are more liable to become depressed than older patients (Richards 1973, Plumb and Holland 1977). Depression in male patients may be more directly related to the consequences of illness, particularly immobility, than in female patients (Roth and Kay 1956). There is a large body of literature associating previous personality characteristics with specific disease states, but rather fewer correlating previous personality with the development of depression in the physically ill. Some investigators have found a higher incidence of prior depressive episodes in these patients, but other groups have failed to confirm this (Roth and Kay 1956, Rimon 1969). Most agree that unlike psychiatric patients with depression there is little evidence of increased frequency of depression in the families of these patients (Rimon 1969). A few studies have mentioned dependency concerns, fears of being 'a burden on others' as being associated with depression (Wright and Owen 1976, Robinson *et al.* 1977). Several investigators have commented on the apparent inverse relationship between the use of denial and depression (Plumb and Holland 1977). Patients who can successfully deny the severity or likely outcome of their illness, at least in the short term, appear to protect themselves from dysphoric mood.

Many studies have mentioned the importance of concurrent life stress, particularly that characterized as object loss in the genesis of depression in the physically ill, much as it is in the case of the non-physically-ill populations (Shochet *et al.* 1969). The social-support system of the patients is receiving increasing attention. Lack of social support has been reported as a factor in the production of depression, while conversely the presence of a good social-support system appears to be an important mitigating force (Cobb 1976, Robinson *et al.* 1977). Socio-economic class by itself is mentioned sometimes, but only as it impinges on the other factors.

References

ALLEN, R. E., and PITTS, F. N., 1978, ECT for depressed patients with lupus erythematosus. *American Journal of Psychiatry,* **135** (3), 367–368.

BARTROP, R. W., LAZARUS, L., LUCKHURST, E., *et al.,* 1977, Depressed lymphocyte function after bereavement. *The Lancet,* **1,** 834–836.

BLACKWELL, B., 1976, Treatment adherence. *The British Journal of Psychiatry,* **129,** 513–531.

CARROLL, B. J., 1978, Neuroendocrine function in psychiatric disorders. In *Pharmacology: A Generation of Progress,* edited by M. Lipton, A. D. Miscio and K. F. Kiliom (New York: Raven Press), p. 487.

CHEN, C. N., 1979, Sleep, depression, and antidepressants. *The British Journal of Psychiatry,* **135,** 385–402.

CLAYTON, P. J., 1979, The sequelae and nonsequelae of conjugal bereavement. *American Journal of Psychiatry,* **136** (12), 1530–1534.

COBB, S., 1976, Social support as a moderator of life stress. *Psychosomatic Medicine,* **38,** 300–314.

CRAIG, T. J., and ABELOFF, M. D., 1974, Psychiatric symptomatology among hospitalized cancer patients. *American Journal of Psychiatry,* **131** (12), 1323–1327.

CRAIG, T. J., and VAN NATTA, P. A., 1979, Influence of demographic characteristics on two measures of depressive symptoms: the relation of prevalence and persistence of symptoms with sex, age, education, and marital status. *Archives of General Psychiatry,* **36,** 149–154.

ENGEL, G. L., 1968, A life setting conducive to illness: the giving up–given up complex. *Bulletin of the Menninger Clinic,* **32,** 355–365.

ENGEL, G. L., and SCHMALE, A. H., 1967, Psychoanalytic theory of somatic disorder: conversion, specificity, and the disease onset situation. *Journal of American Psychoanalytic Association,* **15,** 344–365.

FORESTER, B. M., KORNFELD, D. S., and FLEISS, J., 1978, Psychiatric aspects of radiotherapy. *American Journal of Psychiatry,* **135** (8), 960–963.

GANZ, V. H., GURLAND, B. J., DEMING, W. E., and FISHER, B., 1972, The study of the psychiatric symptoms of systemic lupus erythematosus: a biometric study. *Psychosomatic Medicine,* **34** (3), 207–220.

GOLDBERG, E. L., COMSTOCK, G. W., and HORNSTRA, R. K., 1979, Depressed mood and subsequent physical illness. *American Journal of Psychiatry,* **136** (4B), 530–534.

GREER, S., 1979, Psychological consequences of cancer. *The Practitioner,* **222,** 173–178.

HENDRIE, H. C., 1978, Organic brain disorders: classification, the 'symptomatic' psychoses, misdiagnosis. *The Psychiatric Clinics of North America,* **1** (1), 10.

KAGAN, A. R., and LEVI, L., 1971, Health and environment—psychosocial stimuli: a review. *Reports from The Laboratory for Clinical Stress Research, Number 27.*

KERR, T. A., SHAPIRA, K., and ROTH, M., 1969, The relationship between premature death and affective disorders. *The British Journal of Psychiatry,* **115,** 1277–1282.

LIPOWSKI, Z. J., 1977, Psychosomatic medicine in the seventies: an overview. *American Journal of Psychiatry,* **134** (3), 233–244.

MOFFIC, H. S., and PAYKEL, E. S., 1975, Depression in medical inpatients. *The British Journal of Psychiatry,* **126,** 346–353.

MOLODOFSKY, H., and CHESTER, W. J., 1970, Pain and mood patterns in patients with rheumatoid arthritis: a prospective study. *Psychosomatic Medicine,* **32,** 309–318.

MURPHY, E., and BROWN, G. W., 1980, Life events, psychiatric disturbances and physical illness. *The British Journal of Psychiatry,* **136,** 326–338.

PARKES, C. M., 1970, The first year of bereavement: a longitudinal study of the reaction of London widows to the death of their husbands. *Psychiatry,* **33,** 444–467.

PLUMB, M .M., and HOLLAND, J., 1977, Comparative studies of psychological function in patients with advanced cancer: self-reported depressive symptoms. *Psychosomatic Medicine,* **39** (4), 264–276.

RICHARDS, D. H., 1973, Depression after hysterectomy. *The Lancet,* **2,** 430–433.

RIMON, R., 1969, A psychosomatic approach to rheumatoid arthritis. *ACTA Rheumatogia Scandinovica, Supplement,* **13,** 1–115.

RIMON, R., 1974, Depression in rheumatoid arthritis. *Annals of Clinical Research,* **6,** 171–175.

ROBINSON, E. T., HERNANDEZ, L. A., DICK, W. C., and BUCHANAN, W. W., 1977, Depression in rheumatoid arthritis. *Journal of the Royal College of General Practitioners,* **27,** 423–427.

ROGERS, M. P., DUBEY, D., and REICH, P., 1979, The influence of the psyche and the brain on immunity and disease susceptibility: a critical review. *Psychosomatic Medicine,* **41** (2), 147–164.

ROGERS, M. P., REICH, P., KELLY, M. J., and LIANG, M. H., 1980, Psychiatric consultation among hospitalized arthritis patients. *General Hospital Psychiatry,* **2,** 89–94.

ROTH, M., and KAY, D. W. K., 1956, Affective disorder arising in the senium: physical disability as an aetiological factor. *Journal of Mental Science,* **102,** 141–150.

SACHAR, E. J., HELLMAN, L., ROFFWARG, H. P., *et al.,* 1973, Disrupted 24-hour patterns of cortisol secretion in psychotic depression. *Archives of General Psychiatry,* **28,** 19–24.

SCHMALE, A. H., 1958, Relationship of separation and depression to disease. *Psychosomatic Medicine,* **20** (4), 259–277.

SCOTT, W. A. M., 1969, The relief of pain with an antidepressant in arthritis. *The Practitioner,* **202,** 802–807.

SHOCHET, B. R., LISANSKY, E. T., SHUBART, A., FIOCCO, V., KURLAND, S., and POPE, M., 1969, A medical-psychiatric study of patients with rheumatoid arthritis. *Psychosomatics,* **10,** 271–279.

THORPE, P., and MARCHANT-WILLIAMS, R., 1974, The role of an antidepressant, dibenzepin (noveril) in the relief of pain in chronic arthritic states. *Medical Journal of Australia,* **1,** 264–266.

TSUANG, M. T., and WOOLSON, R. F., 1978, Excess mortality in schizophrenia and affective disorders. *Archives of General Psychiatry,* **35** (10), 1181–1185.

WHITLOCK, F. A., and SISKIND, M., 1979, Depression and cancer: a follow-up study. *Psychological Medicine,* **9**, 747–752.

WIGDOR, B. T., and MORRIS, G., 1977, A comparison of twenty-year medical histories of individuals with depressive and paranoid states: a preliminary note. *Journal of Gerontology,* **32** (2), 160–163.

WRIGHT, V., and OWEN, S., 1976, The effect of rheumatoid arthritis on the social situation of housewives. *Rheumatology and Rehabilitation,* **15**, 156–160.

WRIGHT, V., WALKER, W. C., and WOOD, E. A. M., 1963, Nialamide as a 'steroid sparing' agent in the treatment of rheumatoid arthritis. *Annals of Rheumatoid Disease,* **22**, 348.

ZAPHIROPOULOS, G., and BURRY, H. C., 1974, Depression in rheumatoid disease. *Annals of Rheumatoid Disease,* **33**, 132–135.

Aspects of physiological factors in paced physical work

By E. Kamon

Noll Laboratory,
The Pennsylvania State University,
University Park, PA 16802, USA

The capacity to perform physical work depends on age, gender, and fibre types of the muscles. Therefore, muscular performance is reviewed with respect to maximal performance capacity for dynamic work and for static work on the basis of the expected maximal 0_2 uptake ($\dot{V}0_2$ max) and maximal voluntary contraction (MVC), respectively. Limitation in performance of dynamic and static work due to the muscles' fibre composition is also shown. Not considering fibre types, predictors for working and resting periods, applicable to machine pacing, are given for dynamic and for static work. The schedules of work and rest are aimed at non-fatiguing working conditions and they are based on estimates of the demand of the task and the expected maximal capacity adjusted for age and for gender.

1. Introduction

Since stress–strain relationships in physical work depend on the types of muscular contraction, machine pacing and work–rest schedules should be based on the nature of the contractions. Muscular contractions are broadly categorized as either dynamic or static.

Dynamic effort always involves rhythmical contractions. Static muscular effort could involve either sustained or rhythmical contractions. The physiological responses to the two types of effort are somewhat different. Consequently, the endurance and recovery periods are of different time frames. Overall, the endurance is inversely related to, and the recovery is directly related to, the intensity of the effort. As shown in figure 1a, in both cases the relationships are exponential. Because the limits of dynamic effort is mostly the cardiovascular capacity to meet the 0_2 demand of the muscles, the work can be sustained for prolonged periods. In contrast, static effort occludes the blood flow to the muscles, impairing the blood–muscle exchange of substrates and thus limiting the work to a few minutes or a few seconds.

The demand of a given task is best standardized (for individual differences) in terms of the effort as a fraction of the maximal capacity of the system involved. In dynamic effort it is the maximal capacity for 0_2 delivery and uptake ($\dot{V}0_2$ max per min). In static effort it is the force or the torque yielded

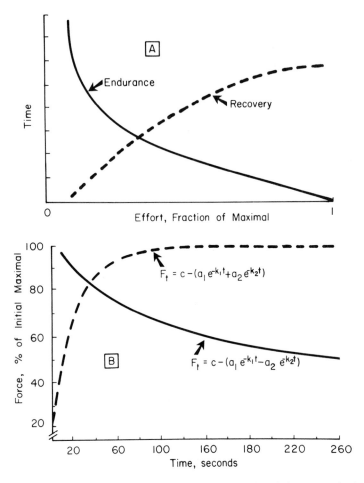

Figure 1. Endurance and recovery: (a) general relationship of time over the intensity of work; (b) strength decrement and strength recovery in rhythmical static effort. Adapted from Clarke (1971) and Stull and Clarke (1971).

during maximal voluntary contraction (MVC) of the muscles involved. At sub-maximal levels of performance, 0_2 uptake as a fraction of $\dot{V}0_2$ max (f $\dot{V}0_2$) and force output as a fraction of MVC (f MVC), best express the expected endurance because these relative values indicate the reserves available to the performer. The closer the values are to maximum, the less is the reserve and thus the shorter the endurance.

The $\dot{V}0_2$ max and the MVC are dependent on the gender, age and fibre composition of the muscles involved. This presentation will focus on these factors as they relate to paced work and to the requirements for schedules of work and rest.

2. Dynamic effort

Dynamic effort refers to the activation of large groups of muscles in tasks such as shovelling, rolling and packing large items, walking, climbing and carrying light weights.

2.1. *Endurance and $\dot{V}0_2$ cost*

The time frame of dynamic effort for the endurance curve in figure 1(a) is as follows: 3–5 min at $\dot{V}0_2$ max; 1–2 hours at $0\cdot5$ $\dot{V}0_2$ max, and 7–8 hours at $0\cdot3$–$0\cdot4$ $\dot{V}0_2$ max. At the same f $\dot{V}0_2$ max fitter persons endure the work longer than less fit persons. However, it is recommended, for safety reasons, to apply the maximal time which is applicable to the less fit worker.

Obviously, the prediction of endurance calls for knowledge of the $\dot{V}0_2$ for the task in question and of the expected $\dot{V}0_2$ max for the potential worker. The $\dot{V}0_2$ cost for a task or for the elements which comprise a given task can be found in Durnin and Passmore (1967) and in Kamon (1981). Some examples of the cost in l min^{-1} are: bagging and packing, $0\cdot5$–$0\cdot7$; shovelling, $1\cdot7$; roof bolting (mines), $2\cdot2$. For lifting, from below knee level, a regression of $\dot{V}0_2$ (ml) on the load moved (kgm) is applicable:

$$\dot{V}0_2 = 600 + 5 \text{ kg m} \tag{1}$$

The expected $\dot{V}0_2$ max for the average person whose task involves the muscles of both the arms and the legs are summarized in table 1. Leg-work alone is expected to elicit 20% higher $\dot{V}0_2$ max and arm-work alone 20% lower $\dot{V}0_2$ max than the values in table 1. Lifting at maximal load (equation 1) either at high rate or by moving heavy weights could elicit $\dot{V}0_2$ max 15% lower than the values in the table (Petrofsky and Lind 1978).

Table 1. Maximal aerobic capacity ($\dot{V}0_2$max) for tasks using arm and leg muscles*.

Age (years)	$\dot{V}0_2$max, l min^{-1}	
	Women	Men
20–29	$2\cdot14 + 0\cdot25$	$3\cdot16 + 0\cdot30$
30–39	$2\cdot00 + 0\cdot23$	$2\cdot88 + 0\cdot28$
40–49	$1\cdot85 + 0\cdot25$	$2\cdot60 + 0\cdot25$
50–59	$1\cdot65 + 0\cdot15$	$2\cdot32 + 0\cdot27$

*Means and S.D. for average fitness level. Adapted from Astrand and Rodahl (1977), Nagle (1973), and Kamon and Pandolf (1972).

2.2. Working periods

Since sound employment practice calls for avoiding undue fatigue, endurance to the limits of performance is undesirable. Indeed, tasks performed at seemingly acceptable $\dot{V}0_2$ were fatiguing, as indicated by the upcreep in heart rate (Hake and Michael 1977, Simonson 1971). The heart rate (HR) and total $\dot{V}0_2$ was used to find the shortest working periods which will prevent fatigue (Astrand et al. 1960, Christensen et al. 1960, Simonson 1971). The data from these observations were used to derive acceptable working periods (Kamon 1978). It was decided to use working time (T_w in minutes) which will be one-third of the time to exhaustion which yielded the formula:

$$T_w = 40 \ (f \ \dot{V}0_2 \ max)^{-1} - 39 \tag{2}$$

2.3. Recovery periods

In general, recovery is exponentially related to the intensity of work (figure 1(a)). In a summary of the available information on HR increments and lactic acid (l.a.) accumulation for short work periods at $0 \cdot 8 \ \dot{V}0_2$ max, lasting $0 \cdot 5$ to $1 \cdot 5$ min, Simonson (1971) concluded that pauses lasting twice as long as the work period could prevent fatigue. The main reason was the transient nature of the very short working periods, which prevented full-scale physiological responses and thus the recovery was sufficient to replenish the substrates used (high-energy phosphate bonds). Such schedules could well apply to design of machine-paced work.

2.4. Resting periods

Steady-state dynamic submaximal work at rates above $0 \cdot 5 \ \dot{V}0_2$ max is accompanied by production of l.a. in the muscles; the amount produced is proportional to the f $\dot{V}0_2$ max. There is a high correlation between the blood level of l.a. and fatigue (Karlsson 1971). Therefore, designing rest periods for steady-state dynamic work can be based on: (1) the time course of l.a. production in the muscles as it appears in the blood; and (2) the rate of l.a. elimination from the blood.

Lactic acid appearance in the blood peaks 4–5 min after the end of muscular work and the rate of elimination is linear, about 3 mg%/min (Stamford et al. 1978). Using the working period at a given intensity (equation 2), and the expected l.a. production and its rate of elimination, provided the following formula for the rest period (T_r in minutes):

$$T_r = 8 \cdot 8 \ ln \ (f \ \dot{V}0_2 \ max - 0 \cdot 5) + 24 \cdot 6 \tag{3}$$

For $\dot{V}0_2$ max, see table 1. This formula applies to work intensity above $0 \cdot 5 \ \dot{V}0_2$ max and is represented in the curve shown in figure 1a.

2.5. *Muscle-fibre composition*

In essence, there are two major types of fibres: slow twitch (ST) and fast twitch (FT). The distribution of the two types of fibres differs from muscle to muscle, and there are individual differences in the fibre distribution for the same muscle. The ST fibres are better users of O_2 than the FT. Therefore, a larger distribution of ST fibres provides for better aerobic endurance of the muscle. Indeed, this was observed when comparing 24 subjects with different fibre distribution in the vastus lateralis (Yates 1980). Since this knee extensor is a prime mover in walking, the $\dot{V}O_2$ max on a treadmill was found to be linearly related to the distribution of the ST fibres (figure 2).

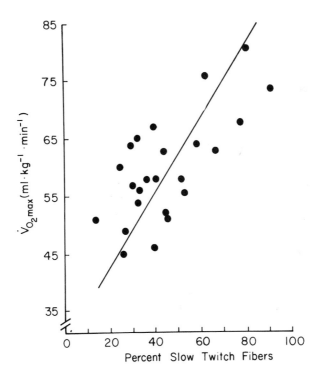

Figure 2. The relationship between maximal O_2 uptake and slow-twitch muscle fibres distribution.

At this state of the art it is premature to include fibre types in the prediction of $\dot{V}O_2$ max or in the design of work–rest schedules. Until predictors for fibre types are established, fibre type is most likely to remain a factor in the large variance seen in the $\dot{V}O_2$ max (table 1).

3. Static effort

Two types of isometric contraction are relevant to occupational stress: sustained and rhythmical. The sustained contraction applies mostly to postural aspects of a job. The rhythmic contraction applies mostly to machine pacing.

The time frame of static effort for the endurance curve (figure 1) is as follows: 5–7 min at $0 \cdot 2$ MVC, 1 min at $0 \cdot 5$ $\dot{V}0_2$ max and $0 \cdot 1$ min at MVC (Rohmert 1960). Because of the impaired blood flow, energy supply for sustained isometric contraction is of anaerobic nature. Since FT fibres are equipped for use of anaerobic processes, during static effort a muscle with a large distribution of FT fibres might have an advantage over a muscle with a small FT distribution.

3.1. *Effect of muscle fibre distribution on static effort*

In the tests described in figure 2 for $\dot{V}0_2$ max and fibre distribution, the subjects were divided according to the FT fibres in their vastus lateralis. Two groups who were distinctly different in FT fibres were selected: (1) fast twitch (FT) with above 60% FT distribution; and (2) slow twitch (ST) with below 40% FT distribution. A static fatiguing exercise followed by a dynamic test was conducted on the two groups using a Cybex isokinetic strength unit (Yates 1980).

Fatigue was developed by static knee extension (knee angle 150°) at $0 \cdot 4$ MVC to exhaustion. The effect of the exhausting exercise was tested with an immediate test of dynamic knee extensions at 11 speeds between 30° s^{-1} to 300° s^{-1}. The torque at knee angle of 150° was measured for each of the dynamic speeds. The results of the torques for each dynamic contraction before and after exhaustion at $0 \cdot 4$ MVC are shown in figure 3. Although before exhaustion the FT group revealed larger torque than the ST at all speeds of contraction, they revealed the same torque as the ST after exhaustion. Thus, FT fibres contribute to more strength, but they appear to fatigue faster than ST. Indeed, at $0 \cdot 4$ MVC, exhaustion occurred on the average in $2 \cdot 7$ min for the FT group and in $3 \cdot 7$ min for the ST group.

3.2. *Sustained static effort*

Endurance for sustained isometric contraction is a function of f MVC. Consequently, the design of work requires knowledge of the torque resisted by the muscles and the maximal torque that the potential worker can exert. While the first can be easily measured, the second can be taken from the expected MVC. Typical maximal torques of the muscles around the primary joints are given in table 2 for average men and women. No age-dependent values are shown. Some observations on men did show decrement of strength due to age (Kamon and Goldfuss 1978), yet others did not (Petrofsky and Lind 1975). Since not enough data are available on the distribution of strength for women and for both genders at advanced age the table is limited to 40 years of age.

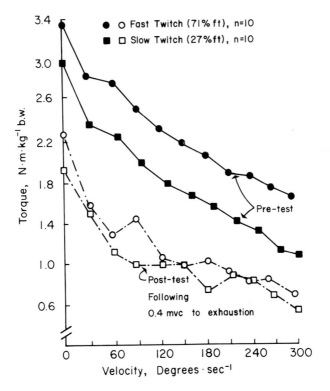

Figure 3. Decrements of dynamic strength following a static effort.

Table 2. The expected maximal torques (N m)† around the joints flexed at different angles for the average men (M) and women (W).

	Torque (N m) at each angle					
	45°		90°		135°	
Joint Action	M	W	M	W	M	W
Shoulder flexion	67	29	68	30	47	21
Elbow flexion	52	24	85	43	60	23
Back extension	—	—	240	130	—	—
Knee extension	135	93	196	130	174	136
Foot plantar flexion	110	83	127	111	101	108

†Mean values (below 40 years of age), with a 25% coefficient of variance. Partially adapted from Laubach (1976).

A formula for static endurance T_w (min) was suggested by Monod and Sherrer (1957):

$$T_w = 0 \cdot 19 \; (f \; MVC)^{-2 \cdot 42} \tag{4}$$

and for recovery was suggested by Rohmert (1960):

$$T_r = 18 \; (t/T_w)^{1 \cdot 4}(f \; MVC \; - 0 \cdot 15)^{0 \cdot 5} \tag{5}$$

where T_r, the resting time in minutes, is a fraction of t the sustained contraction time in minutes. The $0 \cdot 15$ is the asymptotic limit of f MVC for the endurance curve (figure 1).

3.3. *Rhythmical static effort*

This type of muscular contraction is most relevant to machine pacing. A pace of 30 contractions per minute is quite typical. Such a pace was investigated by Clarke (1971) for the decrement of strength and by Stull and Clarke (1971) for the recovery of the strength following the maximal strength decrement. The strength–time relationships are shown in figure 1(b) together with the formulae the authors suggested. It can be seen that there is an asymptotic decrement (c) at $\dot{0} \cdot 5$ MVC. This asymptotic line is rate-dependent: it was found to be $0 \cdot 75$ MVC and $0 \cdot 33$ MVC for paces of 15 min^{-1} and 60 min^{-1}, respectively. Notice the rapid recovery of the strength; more than 60% of the initial MVC is recovered within the first minute.

References

ASTRAND, I., ASTRAND, P. O., CHRISTENSEN, E. H., and HEDMAN, R., 1960, Intermittent muscular work. *Acta Physiologica Scandinavica*, **68**, 448–453.

ASTRAND, P. O., and RODAHL, K., 1977, *A Textbook of Work Physiology* (New York: McGraw Hill).

CHRISTENSEN, E. H., HEDMAN, R., and SALTIN, B., 1960, Intermittent and continuous running. *Acta Physiologica Scandinavica*, **50**, 269–272.

CLARKE, D. H., 1971, The influence on muscular fatigue patterns of the intercontraction rest interval. *Medicine and Science in Sports*, **3**, 83–88.

DURNIN, J. V. G. A., and PASSMORE, R., 1967, *Energy, Work and Leisure* (London: Heinemann).

HAKE, M., and MICHAEL, E., Jr, 1977, The physiological costs of box lifting. *Journal of Human Ergology*, **6**, 167–178.

KAMON, E., 1978, Environmental characteristics. In *Safety in Manual Materials Handling* edited by C. G. Drury (Cincinnati: United States Department of Health, Education and Welfare, Public Health Service, National Institute of Occupational Safety and Health), pp. 155–169.

KAMON, E., 1981, Physiological basis for the design of work and rest. In *Handbook of Industrial Engineering,* edited by G. Salvendy (New York: John Wiley) (in the press).

KAMON, E., and GOLDFUSS, G., 1978, In Plant Evaluation of the Muscle Strength of Workers, *American Industrial Hygiene Association Journal*, **39**, 801–807.

KAMON, E., and PANDOLF, K. B., 1972, Maximal aerobic power during laddermill climbing, uphill running and cycling. *Journal of Applied Physiology*, **32**, 467–473.

KARLSSON, J., 1971, Muscle ATP, CP, and lactate in submaximal and maximal exercise. In *Muscle Metabolism During Exercise*, edited by B. Pernow and B. Saltin (New York: Plenum Press), pp. 289–299.

LAUBACH, L. L., 1976, Comparative muscular strength of men and women: a review of the literature. *Aviation, Space and Environment Medicine*, **47**, 534–542.

MONOD, H., and SHERRER, J., 1957, Capacite de Travail Statique d'um Groupe Musculaire Synergque chez l'Homme. *C. R. Soc. Biol.* (Paris), **151**, 1358–1362.

NAGLE, F. J., 1973, Physiological assessment of maximal performance. In *Exercise and Sports Sciences Reviews Vol. 1*, edited by J. H. Wilmore (New York: Academic Press), pp. 313–338.

PETROFSKY, J. S., and LIND, A. R., 1975, Aging, isometric strength and endurance, and cardiovascular responses to static effort. *Journal of Applied Physiology*, **38**, 91–95.

PETROFSKY, J. S., and LIND, A. R., 1978, Comparison of metabolic and ventilatory responses of men to various lifting tasks and bicycle ergometry. *Journal of Applied Physiology*, **45**, 60–63.

ROHMERT, W., 1960, Ermittlung von Erholungsparesen für Statische Arbeit des Menschen. *Internationale Zeitschrift für Angewandte Physiologie Einschliesslich Arbeitphysiologie*, **18**, 123–164.

SIMONSON, E., 1971, Physiology of work capacity and fatigue (Springfield, Illinois: Charles Thomas Publishers), pp. 440–458.

STAMFORD, B. A., MOFFATT, R. J., WELTMAN, A., MALDONADO, C., and CURTIN, M., 1978, Blood lactate disappearance after supramaximal one-legged exercise. *Journal of Applied Physiology*, **45**, 244–248.

STULL, G. A., and CLARKE, D. H., 1971, Patterns of recovery following isometric and isotonic strength decrement. *Medicine and Science in Sports*, **3**, 135–139.

YATES, J. W., 1980, *The Effects of Muscle Fiber Composition and Static, and Dynamic Exercise on the Torque-Velocity Relationship*, Ph.D. Thesis, The Pennsylvania State University.

The relationship between catecholamine excretion as a measure of psychological stress and the variables of job performance and gender

By C. W. FONTAINE

Digital Equipment Corporation, Maynard, Massachussets, USA

Fifteen males and 15 females worked for five days in a simulated work environment. Physiological changes in the form of greater excretion of both epinephrine and norepinephrine were found, although the increase of norepinephrine was not statistically significant. Performance on the cognitive, data interpretation and utilization, and motor aspects of the complex man–machine task were all found to be facilitated up to a point, but then impaired as the stress level intensified. Perceptual functioning showed no effects of stress. Females, when compared to males, showed a much less severe physiological response to the psychological stress, while performing all tasks with the same efficiency. The results indicate that an inverted-U relationship exists between physiological responses to psychological stress and job performance.

1. Introduction

When stress is viewed as an organism–environment transaction, the variety of situations in which stress may be elicited are numerous, but at no place is the problem more critical than in the work environment (Fröberg et al. 1970). Psychological stress and its associated physiological component have been considered to be primarily products of the social environment; popular and scientific literature has focused on reduction of stress by managerial re-organization, encounter groups, assertiveness training, etc. (Cooper and Payne 1978, Sleight and Cook 1974).

However, psychological stress may also be elicited by the job components themselves. This has been proposed theoretically by Mandler (1976) and verified by Rose (1978). In today's highly technological, mechanized society, many persons work where the possibility of psychological stress is extremely likely.

Unfortunately, very little experimental research has been conducted on occupational stress in which psychological stress has been noted and controlled as an independent variable. The following study was conducted, bringing a real-world problem into the laboratory where it could be examined under controlled conditions. The phenomenon of psychological stress was examined in three basic areas: physiological response, effects on performance and the relationship between the two.

2. Methods

2.1. *Subjects*

The subjects were 15 male and 15 female university students; all had responded to a posted advertisement for temporary employment. The ages of the males ranged from 18 to 27 years (mean = 20·8); females from 18 to 24 (mean = 21·4). The body weights ranged from 63–82 kg (mean = 72·1) in the male group, and 48–68 kg (mean = 57·6) in the female group. All subjects were non-smokers who reported good health at the time of the experiment.

2.2. *Procedure and design*

Subjects received an introductory session in which they were acquainted with the experimental setting and procedures. During this session the subjects were given detailed instructions as to the strict regime which they must follow during the course of the experiment (i.e., restriction of caffeine, alcohol and stress-inducing activities).

Subjects attended each session individually, completing one per day for five consecutive days. In all cases, the first two days were practice; only the last three days were used in data analysis. Each session lasted three hours with a ten-minute break after the first hour. Starting times of the sessions were kept constant for each subject.

The experimental design utilized was a two-factor split-plot design with repeated measures on one variable. In this configuration, the within-subject variable (or repeated measure) was that of stress level (three levels). The between-subjects variable was that of sex (two levels). Stress presentation was counterbalanced.

2.3. *Apparatus and experimental setting*

The experiment was conducted in a large experimental room housed within the psychology department, clearly marked 'Experimental Chambers'. The subject was seated in a swivel armchair behind a small desk nestled between the six experimental chambers in the room. To the immediate left of the subject was a table with two television monitors. To the right was a computer terminal (with a modified keyboard) and its accompanying acoustical coupler and telephone. Also to the subject's right was a 'system failure' box with two levels of failure (Red and Yellow, respectively) and a variety of supporting electrical equipment.

2.4. *Task*

The subjects were instructed to monitor and subsequently control the atmosphere in which experimental animals (rabbits) existed. This notion was supported by an extremely detailed cover story, in which a bogus experiment was outlined.

Within this contrived setting the subjects were asked to perform four embedded tasks. (1) The perceptual task in which the subjects viewed two TV monitors. One displayed an array of meters which the subjects believed indicated the concentration levels of various atmospheric gases found in that specific chamber. On the other monitor, the subjects viewed a rabbit in a cage (supposedly in the chamber in question). In both cases, what the subject saw on the monitors was video-tape previously recorded by the experimenter. (2) The cognitive task involved processing information and performing mathematical calculations according to instructions. (3) The data interpretation and utilization task took the product yielded by the cognitive processing step and applied a set criterion. (4) The final task was that of motor functioning (manually entering required changes into a computer terminal). This, in turn, supposedly adjusted gas valves to give the required atmosphere conditions. In all tasks, raw data was recorded in number of correct responses.

2.5. *Experimental setting*

The independent variable of stress level on the job was manipulated by the alteration of the circumstances under which the subjects performed the job. There were three stress conditions, under which the job was performed: (1) the control condition; (2) normal job stress; and (3) extreme job stress.

In the control condition subjects were told at the outset that a minor problem had arisen: the job would run as normal, but no animals would be in the chambers. Hence, the subjects monitored only the meters. In the normal stress condition the subjects viewed the rabbits in their normal states: alert, hopping, sniffing, etc. Extreme stress, the last condition, was created by introducing an emergency situation into the normal session: the monitor showed the rabbits progressing from a normal state into drowsiness and finally lapsing into unconsciousness. A corresponding change was reflected in the meter readings.

2.6. *Psychoendocrine measurement*

The subjects reported directly to the experimental room, where they received daily instructions and began work. After one hour of work, the subject was given a ten-minute break, during which the subject voided. After the final two hours, subjects gave required urine samples. Each sample was adjusted to a pH of $3 \cdot 0$ with 2M HCl for analysis of free epinephrine and norepinephrine by the fluorophotometric technique (Anderrson *et al.* 1974).

3. Results

Separate analyses were performed for the physiological and the performance measures. In all cases the statistical analyses were executed using the Statistical Analysis System (SAS) (Barr *et al.* 1979) on an IBM 360 computer.

3.1. *Physiological measures*

Epinephrine. The ANOVA performed on the epinephrine measures revealed statistical significance for main effects of both stress and sex. The test for stress for all 30 subjects was found to be highly significant (2 and 56 df, $F = 83 \cdot 17$, $p \langle 0 \cdot 04$). The interaction of stress and sex failed to reach statistical significance. Based on the statistical analysis and subsequent plot of the data, several general statements can be made with respect to epinephrine excretion: (1) a slight difference between the sexes existed in the control condition; (2) a staircase effect was noted for both males and females, where higher epinephrine excretion was noted in response to greater stress; and (3) the increase in the epinephrine excretion noted during stress was more marked for males.

Norepinephrine. The same ANOVA procedure was conducted for the norepinephrine variable. This analysis showed that the main effects of stress and sex, as well as the interaction, failed to reach statistical significance.

3.2. *Task performance*

Multivariate analysis of variance. A multivariate analysis of variance (MANOVA) was performed where all four subtasks were combined and analysed. The results of this analysis clearly showed a highly significant stress effect (10 and 38 df, $F = 7 \cdot 10$, $p \langle 0 \cdot 0001$); but the sex, and the sex and stress interaction, proved to be statistically nonsignificant.

Univariate tests. The results of the ANOVA for the perceptual task showed no statistical signifiance for the effects of stress, sex, or the stress and sex interaction.

The ANOVA for the cognitive task subcomponent revealed a statistically significant main effect: stress affected cognitive performance to a degree which was beyond chance (2 and 56 df, $F = 3 \cdot 64$, $p \langle 0 \cdot 05$). The other main effects, sex and the stress and sex interaction, however, revealed no significance.

The next task performed was that of data interpretation and utilization. Here again, a split-plot ANOVA was performed, and the main effect of stress was found to be significant (2 and 56 df, $F = 34 \cdot 61$, $p \langle 0 \cdot 001$). The main effect of sex and the interaction both failed to reach the $0 \cdot 05$ significance level.

A split-plot ANOVA was also performed on the motor-task performance raw data. The analysis again found a high statistically significant effect of stress (2 and 56 df, $F = 9 \cdot 64$, $p \langle 0 \cdot 001$); while the main effect of sex, as well as the interaction, was non-significant.

Thus far, it has been noted that in three of the four tasks, stress affected the task efficiency to differing degrees, with one task (perceptual) failing to show any performance decrement. Owing to the varying degrees to which stress affected performance on various tasks, general statements are difficult to make. Therefore, an analysis of overall performance was made.

The results of this ANOVA revealed that the main effect of stress was highly significant (2 and 56 df, F = 20·06, $p \langle 0 \cdot 001$). Statistical significance was not reached for the instances of sex, and sex and stress interaction.

3.3. *The relationship between physiological stress responses and performance*

In order to define a general relationship between epinephrine excretion and performance, and to determine if trends noticed in the data were indeed statistically significant, an analysis of variance and subsequent trend analysis on the sums of squares of conditions was conducted. These determined that both epinephrine and overall performance showed stress effects. More

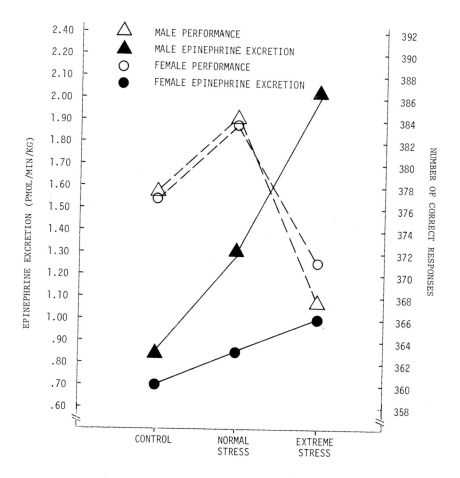

Figure 1. The relationship between mean epinephrine excretion and mean overall performance, for both males and females, as a function of the experimental stress conditions.

important still was the result which showed that epinephrine follows a significant linear trend (1 and 89 df, $F = 9 \cdot 62$, $p \langle 0 \cdot 01$).

The curves for the mean amount of epinephrine excreted as well as overall performance are plotted in figure 1, clearly showing the relationship between epinephrine excretion and performance on the job as a function of increasing levels of stress. Performance increases steadily up to an optimal point (normal stress), after which it begins to decrease. The resulting quadratic trend was highly significant (1 and 89 df, $F = 89 \cdot 81$, $p \langle 0 \cdot 01$).

4. Discussion

The results reveal some striking effects of working under conditions of psychological stress. Firstly, psychological stress may be induced on the job by factors other than those of overt physical threat or interpersonal contact (e.g. coworkers or supervisors). While these factors are sufficient to elicit psychological stress in the workplace, actual components of the job itself may also cause psychological stress, and with it, performance changes—especially where job and/or one's performance challenge one's conception of personal worth or morality.

Secondly, the effects of psychological stress on one's physiological system were quite clear: epinephrine excretion is increased with the introduction of a psychologically stressful component. This epinephrine excretion was found to be highly significant, while it was determined that stressful work situations failed to elicit a significant increase in norepinephrine excretion. These findings confirm the results of Frankenhaeuser (1965).

Interesting results were found concerning the general relationship between physiological arousal and performance. Physiological changes exhibited a linear relationship to the intensity of stress, whereas performance displayed a quadratic relationship to the stress levels. This evidence confirms the notion that an inverted-U relationship exists between job stress and job performance.

Possible sex differences were also investigated. The ANOVA performed on epinephrine excretion found a significant difference between the sexes in epinephrine excretion: females showed less response than males. The magnitude of the difference became even clearer with the comparison of one-way ANOVAS for males and females. Male epinephrine excretion was highly significant ($F = 51 \cdot 40$, $p \langle 0 \cdot 001$), with a statistical difference ($p \langle 0 \cdot 01$) existing between all conditions of stress. In females, while the overall excretion was determined to be significant ($p \langle 0 \cdot 01$), it required the difference between the polar extremes of control and extreme stress to reach that level. No statistically significant differences were found between the middle condition of normal stress or the other two conditions. Whereas both sexes respond physiologically to a stressful situation, it seems that females do so in a much less severe manner. Recalling performance results, it seems that while both sexes are physiologically affected by stress, females are much less so, while their performance remains strikingly similar to that of males (figure 1). Perhaps females reach hyperarousal (and hence, show the observed performance decrement) with substantially lower levels of physiological activation. Such an explanation remains highly conjectural until research on possible sex differences in physiological response to stimuli is conducted.

References

ANDERRSON, B., HOUMOLLER, S., KARLSON, C. G., and SVENSON, S., 1974, Analysis of urinary catecholamines: An improved auto-analyzer fluoresence method. *Report from the Laboratory for Clinical Stress Research,* **32**, 1–36.

BARR, A. J., GOODNIGHT, J. H., SALL, J. P., BLAIR, W. H., and CHILKO, D. M., 1979, *SAS User's Guide* (Raleigh, N.C.: SAS Institute, Inc.).

COOPER, C. L. and PAYNE, R., 1978, *Stress at Work* (New York: John Wiley).

FRANKENHAEUSER, M., 1965, *Physiological, behavioral, and subjective reactions to stress.* Paper presented to the Second International Symposium on Man in Space (Paris).

FRÖBERG, J. E., KARLSSON, D. G., LEVI, L., LIDBERG, L. and SEEMAN, K., 1970, Conditions of work: Psychological and endocrine stress reactions. *Archives of Environmental Health,* **21**, 798–796.

MANDLER, G., 1976, *Mind and Emotion* (New York: John Wiley).

ROSE, R. M., 1978, Health changes in air traffic controllers. *Psychosomatic Medicine,* **40** (2), 142–165.

SLEIGHT, R. B., and COOK, K. G., 1974, *Problems in occupational safety and health: A critical review of select worker physical and psychological factors.* NIOSH Report, MHS-99-82-043 (Cincinnati, Ohio: Department of Health, Education and Welfare).

Section 4.   Measurement of stress

Mental strain in computerized and traditional text preparation

By R. Kalimo and A. Leppänen

Institute of Occupational Health, Helsinki, Finland

Mental strain is known to depend on work characteristics. These in turn are related to technological applications in the work involved. The aim of this study was to investigate the work characteristics, job satisfaction and daily mental strain of workers using traditional methods and computerized techniques with visual display units (VDUs) in text preparation in the printing industry. The subjects were 218 VDU typesetters, perforator type-setters, photocompositors and proofreaders. A questionnaire was administered to all of the subjects. Critical flicker fusion, the Bourdon Wiersma test and a self-rating method were used with different subgroups of the entire sample. The results revealed more positive work characteristics, a higher level of job satisfaction and a lower level of daily mental strain in the groups applying computer technology. The results indicate that the computerization of work processes may increase their positive character if it is applied successfully.

1. Introduction

Mental strain in work has repeatedly been found to depend, among other things, on work characteristics such as variability, autonomy, challenge, and work methods. These are, in turn, substantially related to the technological applications involved in the work. Knowledge based on the study of the relative importance of new technology as such and of the various work designs possible within the technological limitations is definitely needed to promote the job satisfaction, effectiveness and wellbeing of employees.

Studies of mental load in tasks consisting of the processing of numerical and alphabetical data by means of visual display units (VDUs) are scarce. One recent study (Johansson and Aronsson 1979) revealed that the level of mental strain in work with VDUs is essentially dependent on the contents of the tasks. Another study (Wisner 1978), in which two tasks varying in complexity were compared, also revealed higher mental strain in the task that demanded less mental activity than in the more complex task.

A large project, of which this report covers a part, was carried out at the Finnish Institute of Occupational Health to investigate the interrelationships of work characteristics, production technology and the mental and physical strain of the workers (Kalimo *et al.* 1981). The aim of the phase, reported here, was to study the work characteristics, job satisfaction and daily mental strain of groups working in traditional tasks of text preparation and in tasks representing modern applications of computer technology.

2. Methods

2.1. *Subjects*

The subjects of the study were 218 employees in the preprinting text-preparation section of 12 printing plants. They were divided into the following four comparative groups on the basis of a work analysis and the technological equipment in use (table 1).

Group 1 (30 VDU typesetters) transformed manuscript texts into a form that could be further processed. The tool of this group was a machine that resembled a typewriter. It also contained a VDU which the worker used to control performance.

Group 2 (76 perforator typesetters) transferred manuscript texts to a perforated tape which could be further processed.

Group 3 (52 photocompositors) further processed the text set by the typesetters. They worked with a photocomposition machine which included a VDU.

Group 4 (60 proofreaders) read the composed text to find and correct possible errors. Pencils were their only equipment.

Mean ages of the groups varied from 30·5 to 36·4 years (table 1). The groups can be considered comparable in this respect. The sex distributions of the groups varied considerably. Ninety-five per cent of both groups of typesetters were women, 60% of the proofreaders were women, and 75% of the photocompositors were men.

Table 1. Subjects of study.

Technological level	Task	Questionnaire study N	Individual measurements		
			N	Age Mean	Years S.D.	Sex
High	VDU typesetter	30	26	31	8	F
Medium	Perforator typesetter	76	24	31	8	F
High	Photocompositor	52	30	36	11	M
Low	Proofreader	60	19	35	12	F

2.2. *Data collection*

Questionnaire. The subjects' perception of the work characteristics was assessed with a questionnaire. The items formed six scales describing challenge, i.e., mental activity at work, self-determination of the work, role ambiguity at work, social contacts at work, supervision, and appreciation of work. The level of job satisfaction was also measured. The scales consisted of different numbers of items (table 2).

Table 2. Questionnaire scales.

Scale	Number of items	Internal consistency
Challenge	7	0·85
Autonomy	4	0·69
Role ambiguity	6	0·74
Social contacts	4	0·70
Supervision	3	0·71
Job appreciation	4	0·72
Job satisfaction	3	0·78

All of the subjects participated in the questionnaire study, which was administered during a work shift and supervised by a member of the research team.

Measurement of daily stress. Of the total sample, 99 subjects participated in individual tests and measurements reflecting daily workload and stress. Fatigue and alertness were determined from critical flicker fusion (CFF) measurements. Rapidity and exactness of visual perception, combined with motor function, was measured with the Bourdon–Wiersma test. The perceived state was measured with self-rating method. It included 23 items, of which four scales were formed on the basis of a factor analysis (table 3). The measurements were made 1–3 times during a work shift. The subjects were trained in the measurements of CFF two days earlier to avoid the possibility of the measurement being affected by unfamiliarity. Three threshold values, of which a mean value was calculated, were measured three times during a work shift.

Table 3. Self-rating scales.

Scale	Number of items	Internal consistency
Competence	7	0·89
Calmness	4	0·83
Pleasantness	5	0·71
Activation	6	0·86

3. Results

3.1. *Perceived work characteristics and job satisfaction*

The groups' perceptions of the characteristics of their work are presented in figure 1. The challenge experienced deviated between the groups on a statistic-

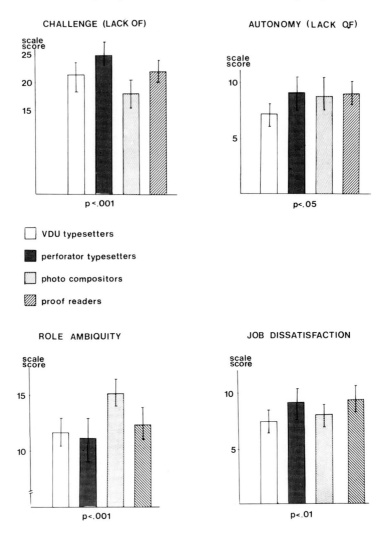

Figure 1. Perceived challenge, autonomy, role ambiguity, and job satisfaction of VDU typesetters, perforator typesetters, photocompositors, and proofreaders.

ally significant level. The photocompositors reported the highest mental activity at work, and perforator typesetters the lowest.

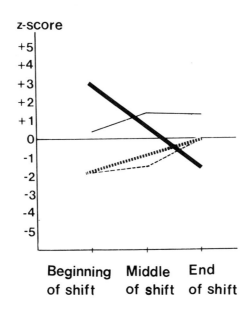

.... VDU typesetters   N=26
▬▬▬ Perforator typesetters   24
- - - - - Proofreaders   19
_____ Photocompositors   30

Critical Flicker Fusion (CFF)

Figure 2. Critical flicker fusion threshold of VDU typesetters, perforator typesetters, photocompositors and proofreaders in the beginning, in the middle, and at the end of the work shift.

The scale for the assessment of autonomy or self-determination at work contained items dealing with the possibilities of moving freely, communication, and taking breaks. The VDU typesetters perceived the highest level of self-determination, and perforator typesetters the lowest.

The level of work-role ambiguity was measured by items concerning the clarity of tasks, instructions and rules. The photocompositors perceived more ambiguity in their roles than the other workers.

No statistical differences between the groups were found for other measured work characteristics.

Job satisfaction was highest in the group of VDU typesetters. The differences between the groups were statistically significant.

3.2. *Daily mental stress*

The three types of data acquired on daily mental stress were standardized so that a comparison between the groups would be possible. Z-scores were calculated from all three types of data.

The changes in the perception of the thresholds of the flickering light in the CFF measurements were small among all the workers except the perforator typesetters. Their CFF threshold declined during the workday on a statistically significant level. This finding indicates an increased level of fatigue during the work shift (figure 2).

The self-ratings of the perforator typesetters differed significantly from those of the other workers. They rated themselves as more fatigued and less competent. Their feelings of pleasantness were also lower (figure 3).

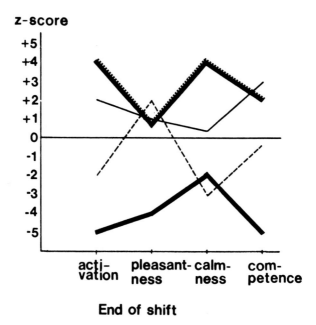

Figure 3. Self-rated state of VDU typesetters, perforator typesetters, photo-compositors and proofreaders at the end of the work shift.

The results of the Bourdon–Wiersma test, which indicated the precision of performance on the basis of visual perception, showed an increasing trend during the day for all of the groups except that of the perforator typesetters. The better results in the repeated testings were due to a learning effect. The perforator typesetters did not show such a learning effect (figure 4).

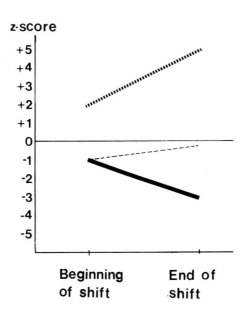

Figure 4. Results of the Bourdon Wiersma test, indicating the accuracy of the visuomotor performance of VDU typesetters, perforator typesetters and proofreaders in the beginning and at the end of the work shift.

4. Discussion

The results indicate that the various work equipment used in the present study can lead to various work organizations, which modify work characteristics and, consequently, may also have an effect on the reactions of the workers. The VDU typesetters and photocompositors, whose tasks represent the most advanced technology applied to text preparation, assessed the central characteristics of wellbeing—mental activity and self-determination at work—more positively than the proofreaders and perforator typesetters.

The differences in the perceived characteristics of the work of the groups of VDU and perforator typesetters are interesting, because the two tasks have the same function in the printing process, i.e., the work includes setting texts from manuscripts. In comparison to the job assessments made by the perforator typesetters, VDU typesetters assess their work as being mentally more demanding and having more self-determination. The VDU typesetters also showed less strain at the end of the work shift. These differences may be caused by many factors. First of all, VDU typesetters can control their own work and receive feedback via the VDU. Perforator typesetters cannot control their work without delay. Working with VDUs is relatively new in the printing industry compared to working with perforators, and this factor may add some fascination and positive attitudes to working with a VDU typesetter. On the other hand, much negative publicity has been directed towards VDUs, a situation which could make attitudes towards that kind of work negative.

The results of the study suggest that tasks involving the computerized preparation of text, which represent the most advanced technological level in the printing industry, may include, at best, favourable characteristics compared to more traditional work methods. The effects of computerization are thus dependent on what is computerized and how the work is organized with new technology.

References

JOHANSSON, G. and ARONSSON, G., 1979, *Stressreaktioner i arbete vid bildskärms-terminal*. Rapport 27,Psykologiska Institutionen, Stockholms Universitet.

KALIMO, R., LEHTONEN, A., SEPPÄLÄ, P., LOUHEVAARA, V., KOSKINEN, P., 1981, *Work arrangement, production technology and mental strain* (Helsinki: Institute of Occupational Health).

WISNER, A., 1978, *Work at computer terminals. Analysis of the work and ergonomic recommendations* (Paris: Laboratoire de Physiologie du Travail et Ergonomie du CNAM).

Physiological monitoring in the workplace†

L. A. GEDDES

Biomedical Engineering Center,
Purdue University, USA

In assessing stress it is customary to select those physiological events that are altered by environmental stimuli. In a laboratory setting the constraints are not too severe. However, in the workplace it is necessary to make a trade-off between those events that are desired and the events that are easily acquired, without hampering the worker in his task. A good principle to employ is 'maximum information from the minimum number of attachments to the subject'. With this as a guideline, a system has been developed to acquire heart rate and its instantaneous change, respiration rate and depth and blood pressure, obtained non-invasively. This system has been used in an industrial setting (Knight *et al.* 1980) to monitor workers on an assembly line.

The method of obtaining heart rate, its beat-by-beat change, repiration rate and depth, requires the application of two small electrodes to the lateral chest wall. One electrode is placed just below the right armpit; the other is placed on the opposite side of the chest along a mid-axillary line, just above the lower border of the rib cage. These two electrodes simulate lead 2 electrocardiogram (ECG) and provide a large amplitude R wave in most subjects. The R wave is bandpass-filtered and fed into a counter to indicate mean or instantaneous heart rate.

Respiration is detected by continuous measurement of the impedance change between the two ECG electrodes. A 25 kHz, constant current (50 μA) is passed through the subject. With respiration, the impedance increases and with expiration, the impedance decreases. The impedance change is processed to provide an analog output which displays rate and depth of breathing (Geddes *et al.* 1962).

Blood pressure is measured with the oscillometric method, which employs a child's cuff placed around the ankle. The cuff is automatically inflated and deflated incrementally, by the Dinamap (Applied Medical Research, Tampa, Florida). From the oscillations imparted to the cuff during deflation, systolic, mean and diastolic pressures are displayed. It is necessary to note that the three

† Supported by NSF Grant APR 77-18695, National Science Foundation, Washington, D.C.

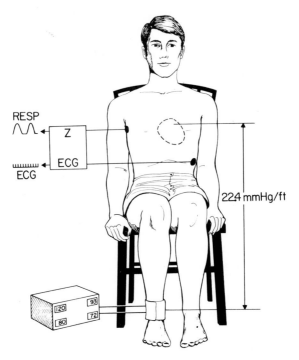

RESP

Z

ECG

ECG

224 mmHg/ft

120   93

80   72

Figure 1. An instrumental subject.

pressures are higher than those measured on the arm by virtue of the pressure due to the column of blood between the heart and ankle. For every foot of vertical displacement between the left ventricle and the cuff, the pressures are elevated by 22·4 mm of Hg.

The analog signals from the acquisition system just described are digitized and recorded on digital tape. The tapes are replayed and programs have been written to permit display of the recorded variables over short or long periods of the work cycle.

References

GEDDES, L. A., HOFF, H. E., HEEKMAN, D. M., HINDS, M., and BAKER, L., 1962, Recording respiration and the electrocardiogram with common electrodes. *Aerospace Medicine*, **33**, 791–793.

KNIGHT, J. L., GEDDES, L. A., and SALVENDY, G., 1980, Continuous unobtrusive performance and physiological monitoring of industrial workers. *Ergonomics*, **23**, 501–506.

Stress:
an analysis of physiological assessment devices

By A. V. KAK

Northeastern Illinois University, USA

1. Introduction

The main focus of this paper will be to evaluate various physiological-dependent measures that have been proposed as indicators of stress. In contrast, the description of particular devices or methodologies for measuring these physiological parameters will only be of secondary importance.

2. Criteria for assessing physiological measures

There are many possible criteria that could be applied for this evaluation but, as an overriding criterion, it is important that the measure be primarily sensitive to *stress* and not to either (1) the stressor itself or (2) the momentary strain of persons' efforts to cope with the stressor. It is crucial to identify measures that adequately separate the contributions of stress, strain and stressor. The set of criteria to evaluate the physiological assessment devices are as follows.

2.1. *Validity*

Two forms of validation are considered, conceptual validity and empirical validity. Conceptual validity refers to the degree of theoretical foundations that supports the connection between a given physiological measure and stress. Empirical validity refers to the degree that a particular measure has been subjected to experimental tests, especially as related to relationships and correlation with measures of strain and stressor level. This particularly includes relationships with subjective measures of stress.

2.2. *Freedom from artefacts*

The criterion of freedom from artefacts refers to the degree to which changes in a physiological measure reflect either (1) other physiological measures unrelated to stress or (2) direct effects of stressors and strain. For example, measures of muscle tension might reflect a component of the stress-response complex but may also reflect momentary physical effort expended to accomplish the demands of a stressor. Similarly, sinus arrhythmia may reflect an aspect of the stress response but may also reflect momentary expenditure of mental effort.

2.3. *Reliability*

Reliability is an obvious criterion and refers to the degree of agreement between repeated measures in the presence of identical stressor levels. Both within subjects and between subjects, consistency is important. Between subjects, reliability may be low both because of difficulty in measuring and assessing a given variable and also because different individuals may show sensitivity to stressors in different ways. Good measures are those which can be counted over a wide range of people.

2.4. *Standardization*

Standardization refers both to the problem of establishing an appropriate non-stress baseline and to interpreting the significance of deviations from this baseline. For example, blood-pressure values are quite well standardized, both in terms of expected population norms and baseline levels as well as in terms of the clinical significance of deviations from baseline.

2.5. *Representativeness*

The criterion of representativeness has two components. First, the stress response has many components and this criterion refers to how well a given single component can serve as a summary for a range of other measures. It is therefore a measure of average correlation between a given single measure and each other measure in a set of stress–response components.

This criterion also refers to how broadly sensitive the measure is to a range of stressor types. Selye (1976) has pointed out that the stress response is non-specific in nature. In other words, the same response may arise from a variety of environmental stressors. If a particular measure is sensitive only in the presence of a particular stressor, it is more reasonable to attribute the effect to some stress-specific action rather than the generalized stress response itself.

2.6. *Invasiveness*

As always, the problem of distorting what is being measured by the very act of measurement is important in assessing stress. Indeed, it has been found that merely displaying a needle to be used for obtaining a blood sample can cause an immediate increase in hormone levels which are intended to serve as stress measures. Ideally, a measure should be obtained unobtrusively. To a large extent, this is a technological problem, since a given physiological measure such as blood pressure may be measured in many ways, some very obtrusive such as direct, invasive recording by pressure transducers, some quite unobtrusive, such as ankle blood-pressure cuffs.

2.7. *Practicality and cost*

The final criteria are related to fairly practical considerations. While primarily technological, they are obviously important. The element of practicality includes a range of factors, including portability and the need for specially trained administrators and technicians, data-analysis techniques and so forth.

3. Evaluation of individual stress-assessment measures

I will now apply these criteria to a range of physiological measures that have been used for stress assessment. Although many other variables have been examined for sensitivity to stress, most of these are not particularly useful as tests of stress in typical industrial situations. For example, various morphological changes in body organs are related to stress but don't lend themselves to use as experimental-dependent measures since, once they occur, they do not readily dissipate with the removal of the stressor.

The evaluation of all the measures is not possible in the space available here. Therefore, I shall present a general overview and highlight some important features of a subset of these measures.

My criteria and measures are organized into the matrix shown in table 1. In this matrix, a ' − ' entry in a cell indicates that the particular measure is especially poor. A ' + ' indicates a special advantage and an entry of "zero" is essentially neutral.

The value in each cell is based upon a review of current literature and practice in this field (Corlett 1973, Mulder *et al.* 1973, Hicks and Wierwille 1979, Hitchen and Harness 1980, Cohen 1980, Selye 1976, Wierwille and

Table 1. Features of physiological measures.

Measures	Concept validity	Empirical validity	Freedom from artefacts	Standard-ization	Reliab-ility	Represen-tativeness	Invasiv-eness	Practi-cality and cost
CFF	−	−	+	−	+	−	0	+
GSR	−	0	−	−	−	0	+	+
EKG	+	+	0	0	+	+	+	0
Heart rate	+	+	0	+	+	+	+	+
Sinus arrhythmia	0	0	0	−	−	0	+	+
Blood pressure	+	+	0	+	+	+	0	+
Pupilary dilation	0	0	−	−	0	−	0	−
Breathing pattern	+	0	−	−	0	0	0	0
Body fluids	+	+	0	+	+	+	+	0
Muscle tension	0	0	0	−	0	0	0	0
EEG	+	0	0	+	0	0	0	0
EP	+	0	0	−	0	0	−	−

Willigres 1980). For an extensive bibiliography, the reader is referred to Wier-wille and Willigres (1980) and Selye (1976).

3.1. *Heart rate*

With respect to mean heart rate, two types of measure should be considered, unadjusted and adjusted or excess. In determining excess heart rate, a basal predicted rate related to oxygen intake is computed according to such models as that of Astrand *et al.* (1973). This is intended to represent the contribution of momentary physical energy demands. This baseline value is then subtracted from observed heart rate and the excess rate is attributed to non-specific and emotional components of the stress response.

Both forms of heart-rate measure have good conceptual validity related both to energy mobilization and to effects of increased blood-hormone levels. With respect to empirical validity, unadjusted heart rate shows a variety of response patterns related to the particular type of stressor. The results with adjusted heart rates seem to be more consistent, generally showing increased excess heart rate in the immediate temporal vicinity of stressor exposure. Blix, *et al.* (1978), for example, using heart rate adjusted for oxygen consumption, reported consistent increases just prior to parachute jumps.

With respect to freedom from artefacts, uncorrected heart rate appears to be a relatively poor measure since it is so sensitive to momentary energy demands, changes in momentary breathing patterns, and so forth. On the other hand, excess heart rate appears to be less susceptible to such errors, since they are essentially accounted for in the computation of adjusted or excess heart rate.

Uncorrected heart rate is well standardized in terms of both the population norms for heart rate and in terms of the physiological consequences of particular values of mean heart rate. However, excess heart rate is far less standardized. The finding of X units of excess heart rate cannot be clearly associated with a particular magnitude of stress across individuals. However, within individuals, excess heart rate and stress are at least ordinally related.

Both forms of heart rate can be reliably measured quite easily. Also, repeated exposure to a given stressful situation seem to reliably produce comparable changes, assuming that artefacts are controlled.

Although both forms of heart-rate measure respond to a wide range of stressors, excess heart rate appears to yield a more consistent, representative response pattern but uncorrected heart rate may show both increases and decreases upon exposure to differing stressors and is also more likely to represent the specific effects of the stressor or strain rather than the stress response itself.

Although both forms of heart-rate measurement can be done with relatively little interference to experimental subjects, accumulating the additional information, such as oxygen consumption, needed to calculate excess heart rate, could greatly increase both the degree of invasiveness or the cost of this measure.

3.2. *Sinus arrhythmia*

Sinus arrhythmia is a relatively new and controversial physiological indicator. It is measured in a variety of ways but is basically the variance in heartbeat to heartbeat intervals observed over a period of time.

This measure has been particularly applied in an attempt to assess mental workload. With respect to conceptual validity, it must currently be given low marks. First, there does not seem to be a clear rationale for why heart-rate variability should decrease as a consequence of mental workload. There are several models which attempt to account for other physiological sources of heart-rate variability such as breathing patterns. However, even if sinus arrhythmia were sensitive to mental workload, this does not necessarily imply a general sensitivity to stressors. In fact, it appears that effects on sinus arrhythmia are very situation-specific, with some types of mental workload producing increases and others showing no effects. In general, although empirical results on sinus arrhythmia are accumulating, it is still too early to see any consistency.

Sinus arrhythmia seems to be quite susceptible to a variety of artefacts. However, many of these influences can be accounted for by the application of physiological models which show the interconnections and feedback loops among breathing patterns, blood volume, blood pressures, mean heart rate and other relevant parameters.

Almost no data on the standardization of sinus arrhythmia is available either for assessing baselines or for assessing the significance of changes in variability. Similarly, the reliability of this measure seems quite poor. There appear to be very large individual differences in the arrhythmia response to a given stressor, and even within a single individual there may be radical variation in the strength and even direction of this measure to repeated applications of the same stress-inducing experimental condition.

The representativeness of this measure is also low. Its magnitude seems to be very closely linked to the specific type of stressor employed. Indeed, one of the supposed advantages of this measure is that it is only sensitive to mental workload, as opposed to physical energy expenditure, for example. Stress responses, on the other hand, are supposed to be similar for a wide range of stressor types. With respect to invasiveness and cost, the measurement of sinus arrhythmia can be given fairly good marks since the data necessary for computing sinus arrhythmia can be recorded as easily as EKG or heart rate.

3.3. *CFF*

Although critical flicker frequency has been used repeatedly in attempts to assess stress, these efforts have not met with success. There is little conceptual validity to this measure and empirical results have failed to reveal a relationship between stress and CFF. This measure seems to be influenced only by stimuli which specifically impact the visual system.

3.4. *GSR*

Like CFF, there is not a good underlying theory to connect changes in skin conductivity or resistance to stress. However, to the extent that GSR reflects resistance decreases caused by sweating, especially palmar sweating, this measure gains more concept validity. There is good empirical support for the validity of GSR as an indicant of stress, with GSR generally decreasing under increasing stress.

Although GSR can be measured quite accurately, it can suffer from several artefacts. In particular, increased palmar sweating may reflect specific stressor effects as well as the stress response itself. Heat as a stressor is an obvious example of a case where a direct stressor effect, rather than the stress response itself, will most likely be measured. Similarly, it is not clear whether tissue conductance changes contribute to the stress response independently of the palmar-sweating effect. However, Katkin has suggested that tissue conductivity may reflect stressor-specific cognitive activity rather than the general stress response. Standardization is also a problem in the use of GSR, since individual differences can be quite large. No clear population baselines are available for this measure, although fairly broad ranges can be specified for expected values of GSR. The baseline value determined for GSR can depend strongly on the distance between measuring electrodes. Thus, changes in GSR are more significant measures. Furthermore, although the appearance of some change in GSR in response to application of a stressor is reliable, the magnitude of this change is far less consistent.

GSR does seem to be fairly representative in the sense that it decreases in response to a wide variety of stressors including heat, pain, cold, induced anxiety about the possible application of painful stimuli, antagonistic interviews and other types.

Finally, GSR can be measured both quite unobtrusively as well as easily and so is rated high on these last two criteria.

3.5. *EEG and EP*

EEG and EP both appear to have reasonable conceptual validity as indicants of stress. Interest has especially focused upon alpha-rhythm blockage as well as changes in other parts of the EEG spectrum. However, this reasonableness of a connection between EEG and stress has not been borne out in empirical results. Although application of certain stressors has resulted in alpha blockage, these effects appear to be quite stressor-specific. At this stage, EEG may serve, at best, as an indicator of arousal. It may also, along with evoked potentials, serve as an indicant of mental workload. Because of the lack of empirical support of this measure, it will not be considered further.

3.6. *Blood pressure*

Blood pressure has proven to be a popular and effective indicator of stress. There is a good physiological theory to give this measure conceptual validity.

This theory relates blood-pressure changes to pituitary and adrenal hormone secretions during stress. Also, this measure is of interest because of its known consequences for long-term health.

A considerable amount of empirical data has been accumulated concerning blood-pressure effects, and the general result is an elevation not particularly sensitive to the specific stressor type. However, under conditions of very severe stress associated with shock, blood pressure generally declines. Also, in the presence of adrenal malfunction, stress may be associated with decreased blood pressure.

Although susceptible to a variety of artefacts, application of various physiological models can be used to remove these to some degree.

Blood pressure is quite well standardized both in terms of its expected baseline values and the consequences of deviations from this baseline on long-term health. It is not well known, however, how small increases in blood pressure over very long time periods might effect long-term health prospects.

A variety of measures are available for reliably measuring blood pressure. Furthermore, this measure produces a fairly good response to the repeated application of the same stressor, assuming that various factors are controlled. One particular difficulty with this measure, especially for field applications, is that it is usually fairly invasive. However, this degree of invasiveness can vary over a large range. Finally, blood pressure can be determined quite easily and practically.

I would like to conclude this brief overview by reiterating what appears to be the main limitation in the application of many of these measures: ensuring that the dependent measure chosen to reflect stress is indeed sensitive to the target response and not to either (1) the actual mental or physical work associated with coping with the stressor or (2) to the direct effects of the stressor itself. It must be emphasized that it is not enough to develop a stress-sensitive measure; equally important is the need to identify and separate the effects of stress, strain and stressor.

References

ASTRAND, I., FUGELL, P., and KARLSSON, C. G., 1973, Energy output and work stress in coastal fishing. *Scandinavian Journal of Clinical Laboratory Investigations*, **31**, 105–113.

COHEN, C. J., 1980, Human circadian rhythm in heart rate response to a maximal exercise stress. *Ergonomics*, **23**, 591–595.

CORLETT, E. N., 1973, Cardiac arrhythmia as a field technique: some comments on a recent symposium. *Ergonomics*, **16**, 3–4.

HICKS, T. G., and WIERWILLE, W. W., 1979, Comparison of five mental workload assessment procedures in a moving-base during similarily. *Human Factors,* **21** (6), 753–761.

HITCHEN, D. A., and HARNESS, J. B., 1980, Cardiac responses to demanding mental load. *Ergonomics*, **23**, 379–385.
MULDER, G., MEULEN, M. H., and VAN DER, 1973, Mental load and the measurement of heart rate variability. *Ergonomics*, **16**, 69–83.
SELYE, H., 1976, Stress in Health and Disease (Sevenoaks: Butterworths).
WIERWILLE, W. W., and WILLIGRES, 1980, *Naval Air Test Center Technical Report, AD A083686.*

Work movements in semi-paced tasks

Institute of Occupational Health, Helsinki, Finland

A review on work movements in semi-paced tasks with special reference to musculoskeletal disorders in repetitive tasks.

1. Strain related to work movements

It has been shown that tasks which include repetitive movements, especially those tasks which are governed by a machine or a production line, are critical from the viewpoint of discomfort, occupational strain and occupational disorders (e.g., Maeda 1977, Grandjean 1979). Work movements—or motor performance—are important to the analysis of strain caused by repetitive tasks, but there are other links in the chain, for instance perception and central control, which cannot be left unanalysed. A central question in this context will be: to what extent do the stresses of a manual task cause strain which may lead to occupational disorders and occupational diseases?

Most of the information gathered to date pertains to tasks which may be called semi-paced. Data about semi-paced tasks has been collected from two sectors: production lines, which allow a certain amount of variation; and services, where the pacing is due to the distribution of clients.

It is evident that strain related to work movements is not the sole cause of occupational disorders in workers with semi-paced tasks. Factors such as individual disposition play an important role, but conclusive evidence about such factors is not as yet available. Figure 1 shows that many primary and secondary factors may be involved.

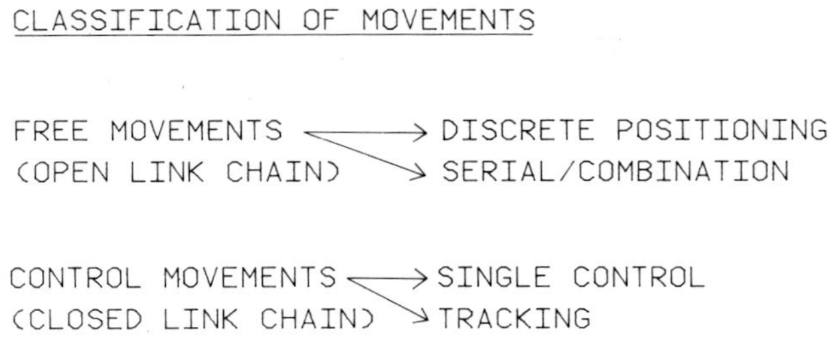

```
CLASSIFICATION OF MOVEMENTS


FREE MOVEMENTS  <───────>  DISCRETE POSITIONING
(OPEN LINK CHAIN)       ↘  SERIAL/COMBINATION


CONTROL MOVEMENTS <────>  SINGLE CONTROL
(CLOSED LINK CHAIN)    ↘  TRACKING
```

Figure 1. Individual factors and task qualities in a repetitive task which may promote the generation of a disorder or an injury.

Whatever the cause may be, strain related to work movements is at least an intervening factor which acts as the medium of different effects. For this reason, work movements must be both analysed and optimized to alleviate strain. It must be emphasized, however, that work movements should not be analysed as a separate entity; such factors as postures, psychological stress and monotony must also be taken into account.

2. Scale of the problem

No comprehensive information about the scale of the problem is available, as there is no standard classification of pathological states, nor is there any uniform system of registration. During the past ten years data has accumulated in various countries. These accumulated sets of data show that the frequency of occupationally related strain can be high in semi-paced types of task. According to one estimation, half of the workers employed in manual, semi-paced tasks suffer more or less from some ailment of the neck region (Luopajärvi *et al.* 1979). Over 10% of these workers experience pain in the hands and arms. It is probable that the problem is grossly underestimated.

3. Classification of work movements

Figure 2 shows a common classification of work movements. Free movements in space characterize the movements in repetitive tasks. Although the physical dimensions of the workplace and of the pieces to be handled set various constraints, there are usually alternative ways of doing the tasks; workers generally make use of these alternatives, e.g., to adapt themselves to the incidents at work (e.g., Wisner *et al.* 1973). Thus, free series of movements

Figure 2. Classification of work movements. Free, serial/combination movements are most commonly encountered in semi-paced tasks.

and combinations of movements are of primary interest. Disorders resulting from various types of control or tracking movements have not been reported.

4. Models for work movements

Guidelines for optimizing work movements have been put forward (e.g., Schulte 1952, Barnes 1968). Although biomechanical aspects are often featured in these guidelines, the guidelines themselves are not based on any specific model of work movements.

Five models can be described. Information models are probably best outlined; the other models either overlap (kinesiologic/biomechanical models) or are general models adapted to work movements (models of control, skill and learning).

4.1. *The kinesiologic model*
The physiology and kinesiology of sports are directed at maximizing the safe and efficient use of one's body. Maximal efficiency is the main objective. Various kinesiologic methods are used towards this end. Kinetic energy used by the different parts of the body when performing has been analysed; the results of these analyses have been used to help maximize efficiency. This approach may prove useful to the analysis of work movements.

4.2. *The biomechanical model*
The biomechanical machinery comprises muscles as the source of power and bones as levers producing force opposing the external forces produced by loads to be moved. In his extensive work on the biomechanics of the upper limbs, Jenik (1973) sets the clarification of optimal conditions which result in effective work movements as the objective of the biomechanics of work movements. He also emphasizes that work physiology must be taken into consideration when the optimal conditions are clarified.

The biomechanical model, in conjunction with knowledge of muscle physiology, has already produced data which can be applied when work movements are planned. Historical knowledge about areas for precise movements at work surfaces, about areas of minimal energy consumption, and about preferable moving directions is still valid (e.g., Murrel 1965).

The static components of work movements constitute an important source of strain. Time-limit curves of static muscle work show that unlimited performance in activities of short duration is possible only if the forces produced are less than 15% of the force produced by maximum effort; as the level of force produced increases, the duration of performance time decreases. Rohmert (1960) and Monod (1972) proposed equations, based on this type of data, which yield the time limit of both intermittent static work and dynamic work. Bouisset (1967) proposed that the maximal speed limit of work movements should be 20% of the possible maximum speed. The most recent contribution was made by Freivalds *et al.* (1979). The biomechanics of work movements provides valuable guidelines by which stress can be reduced when work movements are designed.

4.3. *The information model*

Hick (1952) proposed that, in work movements, the choice reaction time increases with the degree of choice; thus, reaction time was connected to the amount of information to be handled. Hick's law is as follows: $MRT = K \log N$, where N is the number of signals involved in the choice.

Fitts (1954) proposed that the accuracy (i.e., the information) required for a movement determines the time needed for its performance. Later, he developed an index of difficulty (Fitts and Peterson 1964). The index of difficulty is a logarithm function of movement amplitude and the inversion of the width of the target. Several researchers who have used Fitt's principle have, on the whole, found it valid, e.g., Carlton (1980). Kvålseth (1975) studied the effect of a constraint which makes the preview of a movement more difficult. Such constraint related to the movement time more than to the Fitts index of difficulty. Information variables appear to affect movement time more than biomechanical variables.

To date, it is not known if physical strain related to work movements increases according to the number of information variables. Some observations, however, would seem to suggest the possibility. Carlton (1980) investigated the pattern of an aiming movement. He found that error correction near the target increased when Fitt's index of difficulty was high. Wehrkampf and Smith (1952) found that movement time was longer if manipulation at the target was needed. These and other results show that, as the difficulty of movements increases, economical ballistic movements tend to become positioning movements which probably are more straining.

4.4. *The control model*

The upper limb can be simulated as an analogic, technical model of control (Kraiss 1972). To a sufficient extent, the mechanical and control parameters are known to constitute a model. This has been validated in simple movements where the control model has a good fit. The model provides a good idea of the load of the neural control systems in various movements. The usefulness of the model in revealing strains on anatomic structures relevant to disorders remains to be investigated.

4.5. *Skill and learning*

Motor skill and skill acquisition are important aspects to remember when the strain of work movements is considered. As specialists in skill studies point out, the division of skill into perceptual, control and motor aspects is meant more to categorize the subject than to describe the reality. That division is of minor concern to the issue of strain related to work movements, as, in relation to work movements, skills are often dealt with as if they were a black box, a box with learning and individuals as the input, and work performance as the outcome.

Studies on motor skills reveal why work movements may be strenuous. The following draws on Welford's and Singleton's reviews of motor skill (Welford 1976, Singleton 1978). When the subject selects a response, he makes a series of subdecisions. He may have several alternatives or strategies open to him. The strategies employed can be partly dependent on the presentation of information, partly variable at will. The strategy chosen affects the adjacent decisions. If the relationship between signal and response becomes complex, the choice reaction time also increases. This is because the need for a translation mechanism arises, which is probably due to expectations of the coming signal.

Studies have been conducted where the sensory link of the control chain has been disturbed. These studies found that accuracy rather than speed suffers. This finding might have a direct practical value: as Tichauer (1978) points out, thick gloves may disturb the sensations between the fingers, and thereby lead to an insecure grip.

The times of those elements of tasks containing 'perceptual load' seem to be more sensitive to the effects of learning than the movement itself.

The analysis of a skilled performance evidently shows many phases which can be linked to physical strain. Unskilled performance as such is less economical. Semi-paced tasks may also require unexpected skills. In an assembly task, the strategies (i.e., the components of the skills) vary so that incidents of production, changes in the quality of material, fatigue, etc., can be taken into account. Machine-paced tasks seem to require micro-skills which are not registered in any ordinary work analysis.

Besides aspects of control, the general state also affects motor performance. In addition to the effect of fatigue on motor organs, general fatigue clearly increases the control decision times.

The learning of a motor skill provides the operator more reserves: activities of control spread to those work elements which precede and follow the task. In repetitive tasks, the retention and recalling skills required for a certain performance seem to be important.

The analysis of strain related to work movements can benefit from all of the above approaches. However, it would appear that the most relevant information on work movements has been obtained from the study of the biomechanics of the upper limbs. The aspect of skill also seems to provide knowledge which illustrates some of the aspects of strain and the aetiology of disorders in repetitive tasks.

References

BARNES, R. M., 1968, *Motion and Time Study* (New York : John Wiley).

BOUISSET, S., 1967, Postures et mouvements. In *Physiologie du Travail* edited by J. Scherrer (Paris: Masson), Vol. 1. pp. 112-153.

CARLTON, L. G., 1980, Movement control characteristics of aiming responses. *Ergonomics*, **23**, 1019-1032.

FITTS, P. M., 1954, The information capacity of the human motor system in controlling the amplitude of movement. *Journal of Experimental Psychology,* **47**, 381-391.

FITTS, P. M., and PETERSON, J. R., 1964, Information capacity of discrete motor responses. *Journal of Experimental Psychology*, **67**, 103-112.

FREIVALDS, A., LEE, M. W., and CHAFFIN, D. B., 1979, Towards a rest allocation scheme for sequential static muscle exertions. In *Compass for Technology*, Proceedings of the 23rd annual meeting of the Human Factors Society, edited by C. K. Bensel (Boston, Massachusetts) Oct. 29 – Nov. 1, pp. 205-209.

GRANDJEAN, E., 1979, Psychological and physiological reactions associated with repetitive tasks. In *Compass for Technology*, Proceedings of the 23rd annual meeting of the Human Factors Society, edited by C. K. Bensel (Boston, Massachusetts) Oct. 29 – Nov. 1, pp. 21-24.

HICK, W. E., 1952, On the rate of gain of information. *Quarterly Journal of Experimental Psychology*, **4**, 11-26.

JENIK, P., 1973, *Biomechanische Analyse ausgewählter Arbeitsbewegungen des Armes. Schriftenreihe 'Arbeitswissenschaft und Praxis* (Berlin-Köln-Frankfurt: Beuth-Vertrieb), p. 194.

KRAISS, K. -F., 1972, A model for analysing the coordination of manual movements. In *Displays and Controls*, edited by R. K. Bernotat and K. -P. Gärtner (Amsterdam: Swets & Zeitlinger, N. V.), pp. 155-173.

KVÅLSETH, T. O., 1975, A model of linear arm movements with preview constraints. *Ergonomics*, **18**, 529-537.

LUOPAJÄRVI, T., KUORINKA, I., VIROLAINEN, M., and HOLMBERG, M., 1979, Prevalence of tenosynovitis and other injuries of the upper extremities in repetitive work. *Scandinavian Journal of Work Environment and Health*, supplement 3, pp. 48-55.

MAEDA, K., 1977, Occupational cervicobrachial disorder and its causative factors. *Journal of Human Ergology*, **6**, 193-202.

MONOD, H., 1972, How muscles are used in the body. In *The Structure and Function of Muscle*, edited by G. H. Bourne (New York and London: Academic Press), Vol. 1, pp. 24-74.

MURREL, K. F. H., 1965, Ergonomics, Man in His Working Environment (London: Chapman and Hall), p. 496.

ROHMERT, W., 1960, Ermittlung von Erholungspausen für statische Arbeit des Menschen. *Internationale Zeitschrift für Angewandte Physiologie Einschliesslich Arbeitsphysiologie*, **18**, 123-164.

SCHULTE, B., 1952, *Arbeitserleichterung durch Anpassung der Machine an den Menschen* (München: Hanser Verlag), Vol. 1, p. 81.

SINGLETON, W.T., 1978, Laboratory studies of skill. In *The Analysis of Practical Skills*, edited by W. T. Singleton (Lancaster, England: MTP Press), pp. 16-43.

TICHAUER, E. R., 1978, *The Biomechanical Basis of Ergonomics. Anatomy Applied to the Design of Work Situations* (New York, Chichester, Brisbane, Toronto: John Wiley), p. 46.

WEHRKAMPF, R., and SMITH, K. U., 1952, Dimensional analysis of Motion, II. Travel,

distance, effect. *Journal of Applied Psychology*, **36**, 201–206.
WELFORD, A. T., 1976, *Skilled Performance: Perceptual and Motor Skills* (Glenview, Illinois: Scott, Foresman and Co.).
WISNER, A., LAVILLE, A., TEIGER, C., and DURAFFOURG, J., 1973, *Consequences du Travail répétitif sous cadence sur la santé des Travailleurs et les Accidents.* CNAM, Laboratoire de Physiologie du Travail, Rapport No. 29 bis, p. 55.

Human factors methodology for stress evaluation in machine-paced jobs: some NIOSH experiences

A. COHEN

Applied Psychology and Ergonomics Branch,
Division of Biomedical and Behavioral Science,
National Institute for Occupational Safety and Health,
Cincinnati, Ohio 45226, USA

Procedures for examining machine-paced job demands and related work conditions in terms of their stress-producing effects upon worker health and wellbeing are described based on NIOSH research. Among those presented are questionnaire surveys intended to define psychological stressors and strains, on-site observations with checklists to focus on work-station/environmental/equipment/work-process factors of concern, and task analysis and modelling techniques to characterize the physical demands imposed and means for their alleviation.

1. Introduction

Job demands which cannot be met by the worker constitute stress, as do work conditions which deprive the worker of need fulfillment. Determinants of job stress are numerous and varied, encompassing psychological, bio-mechanical and environmental types of factor. Machine-paced work repre-sents only one potential source of job stress which could be either psycho-logical or biomechanical in nature, depending on the mental or physical effort required. Outcomes of job stressors, called strains, also vary widely. They range from negative attitudes and moods through maladaptive behaviours and physiologic dysfunction to ill-health.

Over the years, the National Institute for Occupational Safety and Health (NIOSH) has used different methods to research these assorted job stressors and strains. Included are: (1) questionnaire surveys for depicting psychological job stressors and their adverse effects; (2) on-site observations with checklists to assess environmental/equipment/workstation features of concern; and (3) job-task analysis and modelling to isolate elements of biomechanical stress and to identify possibilities for their reduction. NIOSH experiences in applying these approaches to job-stress issues, especially problems posed by machine-paced work, are the subject of this chapter. As will be seen, these methods are not without limitations and their use in some instances warrants additional measures to buttress results.

2. Questionnaire approaches

Figure 1 charts the variety of stress-producing factors at work, contributing influences, and stress outcomes as found in typical models which relate psychological job stress to worker strains. Questionnaire approaches are best at providing information on these assorted variables, and NIOSH uses such instruments in its studies of stress arising from machine-paced work. In content, the NIOSH questionnaire material draws heavily from pre-existing survey instruments which have been developed for stress evaluation. For example, job stressors relating to workload or work-pressure factors, under-utilization of skills, job-future ambiguity, non-participation in job decisions, and physical discomfort are defined by scales of items taken from the Caplan *et al.* (1975) questionnaire study of job demands and health in 23 different occupations and components of the Work Environment Scale designed by Insel and Moos (1974). Measures of social support, considered a buffer against job stress, also utilize the Caplan *et al.* (1975) scales which refer to aspects of supervisory interaction and relations with coworkers. In terms of outcome measures, psychological-strain indicators such as workload dissatisfaction, boredom and somatic complaints again borrow from the Caplan *et al.* questionnaire form. In addition, items from the Profile of Mood States instrument as developed by McNair *et al.* (1971) are incorporated to depict one's affective state, e.g. tense/anxious, dejected/depressed, angry/hostile. Separate additions to the questionnaire by NIOSH cover demographics, a

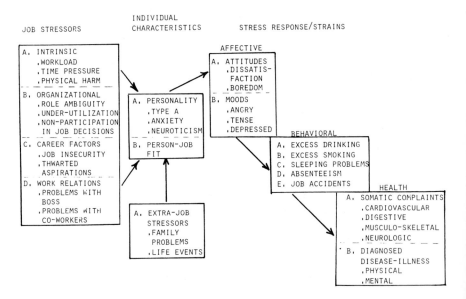

Figure 1. A model relating psychological stressors on the job to consequent worker strains. Adapted from Cooper and Marshall (1976).

checklist to ascertain health problems of possible stress origin, and items indicating behavioural responses to stress (e.g., increased smoking, drinking, sleeping problems, frequent accidents, absenteeism).

Use of this questionnaire form as adapted to specific evaluations of job stress from machine-paced work has not been without its problems.

(1) The form is quite lengthy due to the multiplicity of factors under examination. Completion takes a minimum of 45 min and more likely 60 min. Response rates, given mail-out or on-site distribution to workers, is in the 50% to 60% range even with follow-up contacts and use of token cash rewards for participation. Response rates at this level may raise questions of the representativeness of the study results.

(2) Selection of an appropriate comparison or control group for these questionnaire surveys can be difficult. A group of workers not subject to machine-paced routines but who otherwise work in similar jobs, at the same worksites, and are matched in age, sex and job service are the logical choice. However, this latter group may be experiencing other job pressures which can affect the stress–outcome measures in ways much like those due to machine pacing. Such results can obscure the true level of the particular stressor operating in the target group under study.

(3) Aspects of machine-paced work or other job conditions presumed to be stress-producing (e.g., shift-work regimens) can also be issues in labour–management negotiations. Disputes here can create situations where questionnaire data indicating the stress impact of such factors, being based on subjective responses, can be open to profound bias. NIOSH has attempted to circumvent this problem by replicating the same surveys at other worksites where labour–management relations are less problematic, or by testing and retesting the same worker groups as part of a longitudinal study to confirm the original observations. Ongoing NIOSH studies of stress from machine-paced work in letter sorting, and from use of video display terminals have these add-on features to gain verifiable data.

Even with the above-mentioned limitations, questionnaire approaches for examining job stress and strain due to machine pacing or other job features are still viewed as the only practical way to furnish data on the diverse and subtle factors of relevance. The outputs from such survey procedures are admittedly first-order appraisals. Follow-on work of a more controlled or focused nature to gain a fuller understanding of apparent problems is essential.

3. Site observations and check-lists

In actuality, NIOSH investigations of job stress and strain from machine pacing and other workplace demands are rarely based on questionnaire findings alone. On-site observations of the working conditions found in a sample of the workplaces included in such surveys typically complement the questionnaire results. NIOSH has composed checklists for this purpose, giving attention to physical-environmental factors, workstation layout, equipment and work-process features which can interact with the job demands and intensify the stressors and strains under study.

The combined questionnaire survey and site-observation approach was used successfully in a recent NIOSH evaluation of job stress in video display terminal (VDT) tasks, which included some forced-pace regimens (Smith *et al.* 1981, Murray *et al.* 1981). Visual disturbances (e.g. blurred vision, eyestrain), and musculoskeletal complaints (e.g., neck/shoulder stiffness, back pain, sore wrists), dominated the questionnaire responses of workers engaged in such tasks. These strains could be traced to illumination problems (e.g., background lighting affording poor contrast ratios for VDT viewing, numerous glare sources) and use of office furniture poorly adapted to the more critical viewing/postural demands of the task as found in surveying their worksites.

A rather simple checklist is now being used by NIOSH to obtain some indication of the prevalence of job tasks which are repetitive in nature and could through their recurring physical or mental demands cause health problems, variously defined as 'wear and tear' injury or 'chronic trauma'. NIOSH surveyors have recently begun interview and walk-through surveys that will ultimately cover 5000 establishments, mostly manufacturing in nature, and employing nearly 30 million workers in 98 different sampling districts around the USA. The major purpose of this effort is to note chemicals in use and the presence of harmful physical agents together with preliminary estimates of the number of workers exposed. In addition to these observations, which will provide a prevalence base for characterizing chemical/physical hazards in industry, the surveyors have also been instructed to record certain worker activities which involve frequent repetition and/or significant physical or mental loads. Only ten types of activity are to be noted, in keeping this add-on task simple and not too burdensome to the surveyors. One such category deals with machine-paced operations requiring workers to perform repetitive acts at set times. Other categories include long-term standing (e.g., machine tending), awkward postures (e.g., assuming kneeling, semi-prone positions *re* maintenance jobs), repeated lifting/manual transport of materials, monitoring or watch-keeping. Findings from this survey will provide the first type of surveillance data on the workpacing factor as well as others posing possibly stressful physical and psychological job demands.

4. Task analysis and modelling

Certain jobs requiring repetitive, forceful exertions of the forearm, wrist, hands and fingers have already been linked with carpal tunnel syndrome, tenosynovitis, ganglionic cysts and related musculoskeletal disorders (Hymovich and Lindholm 1966, Mohr 1977). Small-parts assembly, upholstery stitching, and packaging are among those occupations reporting frequent problems of this type (Rabourn 1977, Armstrong *et al.* 1981). NIOSH has been supporting research dealing with the postural elements of these jobs, the biomechanical forces involved, and ways for reducing consequent musculoskeletal strain (Armstrong and Chaffin 1976, Armstrong *et al.* 1981). Methods developed for this purpose consist of filming and frame-by-frame analysis of the worker's arms, wrist, hand, and finger positions while performing the work task, coupled with electromyographic (EMG) recordings to describe the muscle forces involved. The results, presented in terms of frequency distributions of postures and EMG readings, offer a basis for sorting out the more stressful task elements which then become candidates for job redesign. Tool substitution or its repositioning, changes in workstation height or layout are some of the techniques recommended for reducing undue biomechanical forces. Because of its cumbersome detail, this type of analysis is conducted with a few experienced workers who model the job task under study. This could be a weakness in the procedure since the performance data from these select workers is assumed to be characteristic of the total workforce engaged in such tasks. Indeed, Rabourn (1977) found that those workers with a history of carpal tunnel syndrome exert higher average forces in their wrist/hand movements on a job task than those not so afflicted. Nonetheless, expected force reductions through task redesign suggest magnitudes which can exceed these intersubject differences.

Armstrong *et al.* (1981) are currently using this method in a NIOSH contract study of a machine-paced poultry process. Turkey carcasses are hung on a moving conveyor line with attendant workers making specialized cuts to supply various products for the market. Figure 2 shows an oyster cut which is the task having the highest reported frequency of wrist disorders based on the

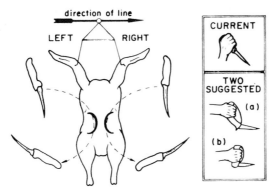

Figure 2. Nature of 'oyster cuts' on turkey carcasses moving on a conveyor line. Current knife and hand grasps are shown along with two suggested designs to reduce wrist deviations. Taken from Armstrong *et al.* (1981).

plant records. An oyster cutter can cut six turkeys per minute, and make from 2500 to 5000 cuts per work shift, depending on the turkey size. Film analysis of postural movements and EMG readings obtained on two workers in this task indicated the presence of sustained grip forces and rapid cycles of extreme wrist flexion and extension. Such biomechanical demands are considered determinants in carpal tunnel syndrome. Redesigned knife handles (inset in figure 2) are among the ideas being evaluated for reducing these stress movements.

In summary, though improvements are needed, each of the methodologies presented here supplies some illumination of the job-stress problem. Perhaps most commendable is the fact that, being applied in the field, they provide for appraisals of the issues as they exist in the real world.

References

ARMSTRONG, T., and CHAFFIN, D., 1976, *An investigation of occupational wrist injuries in women.* NIOSH Grant No. 2 RO1-OH 00679, Center for Ergonomics, University of Michigan, Ann Arbor, Michigan 48109.

ARMSTRONG, T., CHAFFIN, D., and FUOLKE, J., 1979, A methodology for documenting hand positions and forces during manual work. *Journal of Biomechanics,* **12,** 131–133.

ARMSTRONG, T., FOULKE, J., GOLDSTEIN, S., and JOSEPH, B., 1981, *Analysis of cumulative trauma disorders and work methods,* Interim Report to NIOSH, Center for Ergonomics, University of Michigan, Ann Arbor, Michigan 48109.

CAPLAN, R. D., COBB, S., FRENCH, J. R. P., HARRISON, R. V., and PINNEAU, S. R., 1975, *Job demands and worker health.* US Department of Health and Human Services, NIOSH Publication 75-160.

COOPER, G. L., and MARSHALL, J., 1976. Occupational sources of stress: A review of the literature relating coronary heart disease and mental ill health. *Journal of Occupational Psychology,* **49,** 11–28.

HYMOVICH, L., and LINDHOLM, M., 1966, Hand, wrist and forearm injuries: The results of repetitive motions. *Journal of Occupational Medicine,* **8** (11), 575–577.

INSEL, P., and MOOS, R., 1974, *The work environment scale* (Palo Alto: Consulting Psychologist Press).

McNAIR, D. M., LORR, M., and DROPPLEMAN, L. F., 1971, *Profile of Mood States* (San Diego, California: Educational and Industrial Testing Services).

MOHR, E. G., 1977, *Investigation of manual work methods as etiological factors of carpal tunnel syndrome.* Department of Industrial and Operations Engineering, University of Michigan, Ann Arbor, Michigan 48109.

MURRAY, E. E., COX, C., SMITH, M. J., and STAMMERJOHN, L. W., 1981, *Potential health hazards of Video Display Terminals, Technical Assistance Reports of California Investigations* (Cincinnati, Ohio 45226: National Institute for Occupational Safety and Health).

RABOURN, R., 1977, *Investigation of individual work methods as etiological factors of carpal tunnel syndrome*. Center for Ergonomics, University of Michigan, Ann Arbor, Michigan 48109.

SMITH, M. J., COHEN, B. G. F., STAMMERJOHN, L. W., and HAPP, A., 1981, An investigation of health complaints and job stress in video display operations. *Human Factors* (in the press).

Mental strain in machine-paced and self-paced tasks

R. Kalimo, A. Leppänen, M. Verkasalo, A. Peltomaa and P. Seppälä

Institute of Occupational Health, Helsinki, Finland.

The aim of the study was to clarify the effect of machine pacing on mental strain. The subjects were 330 female workers doing repetitive tasks in the printing industry. The subjects represented three forms of workpacing: unmechanized self-pacing, a medium level of machine pacing with one machine, and a high level of machine pacing on a machine line.

A questionnaire, a daily self-rating of one's own state, and cortisol measurements from urine samples revealed only a few differences between the groups. There was a slight indication of higher mental strain among the members of the highly machine-paced group when they were compared to the subjects of the study. The similarity of the responses of all the groups may be due to a selection to and from the occupations under study, adaption, and a relative flexibility of the machine lines studied.

1. Introduction

Machine pacing is a common characteristic of work conditions in various mechanized industrial fields. Forced workpace is often combined with repetitive work content and short operational cycles. These factors often form a combination of quantitative overload and qualitative underload. Underload and overload are not the only problems associated with machine pacing. A standardized workspeed also ignores the varying rhythmicity of psychic and physical functions of man.

Laboratory studies have shown that the amount of physiological strain increases with an increase in machine pacing (e.g. Mackay *et al.* 1979). Field studies have also reported findings which indicate elevated psychic tension and psychophysiological reactions (Johansson *et al.* 1978; review by Cox 1980).

The purpose of this study was to clarify the effect of machine pacing on daily strain, long-term strain and job satisfaction of workers working at various levels of machine pacing in repetitive tasks with low variation.

2. Methods

2.1. *Subjects*

The subjects were 330 female workers in post-printing tasks in 12 printing plants. One group represented work with no machine pacing; the second group

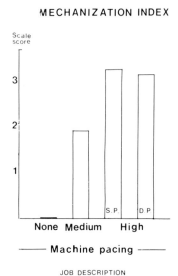

MECHANIZATION INDEX

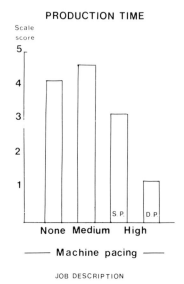

PRODUCTION TIME

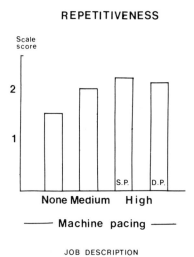

REPETITIVENESS

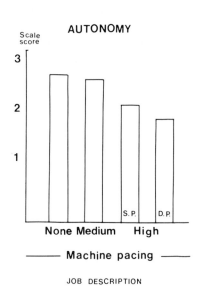

AUTONOMY

Figure 1. Mechanization index, production time, repetitiveness of work, and autonomy of the workers on the basis of job description of the tasks at various levels of machine pacing. S.P. indicates seasonal product and D.P. daily product.

worked with single machines at a medium level of machine pacing; and the third group comprised workers at machine lines (table 1). Task characteristics based on a work description are given in figure 1.

Table 1. Characteristics of the groups studied.

Machine pacing	Technology applied	Type of product	Number of subjects		
			Questionnaire study	Individual measurements	
				Morning shift	Evening shift
None	Unmech-anized	Seasonal (books & magazines)	125	17	11
Medium	Single machine	Seasonal (books & magazines)	83	12	—
High	Machine line	1. Seasonal (books & magazines)	103	30	18
		2. Daily (newspapers)	99	—	10

The mean age ranged from 40·6 to 43·3 years (SD:s from 9 to 12 years). The length of work experience ranged from 12·3 to 17·1 years (SD:s from 9 to 11 years). Education was about the same in all groups. All of the subjects participated in a questionnaire study.

Smaller samples were chosen for individual measurements to determine the daily workload. Selection of the workers was made on the basis of age, health, medication, smoking, etc., because of the hormonal measurements. The day-shift groups were age-matched well, the mean age ranging from 37·9 to 40·0 years. Among the evening-shift samples the mean age ranged from 32·6 to 40·8 years.

2.2. *Collection of data*

Questionnaire. A questionnaire was used to gather information on the following aspects: work characteristics; work environment; job satisfaction; psychic and psychosomatic symptoms; and psychic resources. The questionnaire included also a list of mental abilities and skills; the subjects rated these according to their importance at work. The questionnaires were completed at the workplaces during the work shift.

The questionnaire items measuring work characteristics and workers' reactions were factor-analysed separately, after which multi-item sum scales were formed (table 2). The reliabilities of the scales were between 0·68 and 0·85.

Hormonal excretion. Daily mental load and stress reactions were studied by neuroendocrine measurements and subjective ratings. Of the neuroendocrine reactions, only the excretion of cortisol is reported here. Urine samples were collected at the end of the work shift. These samples were used to determine the excretion level during the last half of the work shift. This period was preceded by a normal break. Control samples were collected during leisure time, at 6 a.m. on the day which followed two weekly free days. The work-shift samples were collected during the same week.

Daily self-rating. A subjective rating of one's own perceived psychic state was carried out at the end of the same work shift in which the urine samples were collected. A semantic differential method with 23 items measuring competence, pleasantness, tension and fatigue was used. The reliabilities of the sum scales formed on the basis of factor analysis varied between $0 \cdot 71$ and $0 \cdot 89$.

3. Results

3.1. *Job demands on mental abilities and skills*
All study groups perceived the 20 work demands on mental abilities and skills about equally. These results support the comparability of the study groups with regard to the demands of ability and skill at work.

3.2. *Perceived work characteristics*
Significant differences between the groups were found in the scales reflecting the challenge and the self-determination of work (table 2). Those working with single machines at a medium level of machine pacing found their work more challenging but less self-determined than the others.

3.3. *Job satisfaction, perceived symptoms, and resources*
The level of general job satisfaction was about the same in all study groups. Among the symptoms, statistically significant differences were found only in sleep disturbances, which occurred most often among machine-line workers, and in escape reactions, which occurred most often among workers in unmechanized tasks. Psychosomatic symptoms were somewhat more common among the workers in highly machine-paced work (table 2).

3.4. *Daily strain*
The levels of the excretion of cortisol at work expressed as percentages of the leisure-time baselines did not differ significantly between the groups in the morning shift (figure 2). There was, however, a slight tendency towards a higher excretion in the highly machine-paced group.

The excretion of cortisol in the evening shift was measured in workers in un-mechanized tasks and in the workers of two groups with the highest machine pacing. One of them represents seasonal products and the other daily papers. This comparison was made to study the possible effects of time pressure,

Table 2. Mean values of perceived work characteristics, job satisfaction and symptoms in the groups studied. A high score indicates a negative situation.

	Unmechanized work N=95–123	Single machine N=67–80	Machine line Seasonal product N=74–99	Machine line Daily product N=75–96	One-way analysis of variance
Work characteristics:					
Challenge (7 items)	26·6	25·0	27·9	26·4	F(3·356) = 5.90***
Self-determination (4 items)	9·3	10·7	10·4	9·4	F(3·379) = 5·67***
Role ambiguity (6 items)	11·9	11·8	11·8	12·2	NS
Social contacts (4 items)	9·1	9·4	9·2	9·0	NS
Supervision (3 items)	8·4	8·7	8·9	8·4	NS
Job appreciation (4 items)	12·9	12·0	12·3	12·1	NS
Job satisfaction (3 items)	8·7	9·0	9·3	8·6	NS
Perceived symptoms:					
Monotony at work (4 items)	11·5	10·8	11·6	10·7	NS
Psychic symptoms (6 items)	17·8	17·4	18·1	17·3	NS
Psychosomatic symptoms (8 items)	17·5	16·4	17·8	18·0	NS
Memory problems (4 items)	9·6	8·9	9·4	9·5	NS
Sleep disturbances (3 items)	6·5	6·1	7·0	7·0	F(3·397) = 2·75*
Escape reactions (2 items)	4·6	4·1	4·5	4·1	F(3·398) = 3·87**
Psychic resources (5 items)	13·4	12·4	13·4	12·7	NS

NS = not significant * = $p < 0·05$ ** = $p < 0·01$ *** = $p < 0·0001$

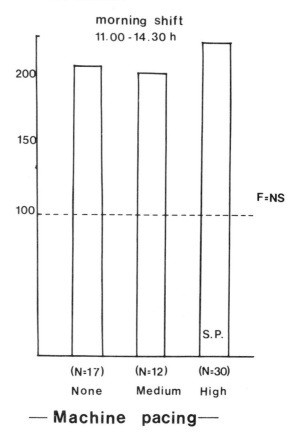

Figure 2. Excretion of cortisol by workers at various levels of machine pacing in the morning shift.

which was expected to be higher at the newspaper lines. No difference was found in the endocrine responses of the groups. There is, instead, a tendency towards a lower excretion in workers producing newspapers. This is probably due only to diurnal variation, as the work shift lasts longer in newspaper production (figure 3).

The daily self-ratings of competence, pleasantness and tension were about the same in all study groups (figure 4). Perceived fatigue was stronger in the workers of the highly machine-paced group producing newspapers, which may be due to the later hours of the work shift.

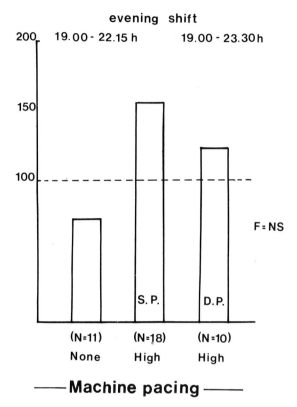

Figure 3. Excretion of cortisol by workers with no machine pacing and with high machine pacing in the evening shift. S.P. indicates seasonal product and D.P. daily product.

4. Discussion

Perceived work characteristics, job satisfaction, symptoms and resources, and daily strain differed only little between the groups working at various levels of machine pacing. There were slight tendencies indicating greater work-load and stress in the workers carrying out machine-paced tasks. Several possible hypotheses can be proposed to explain the findings.

One of the first questions is the possible insensitivity of the methods. This suspicion, however, is not justified, as the methods proved sensitive enough to differentiate the groups now studied from several text preparation groups in the pre-printing phase of the process (Kalimo *et al.* 1981).

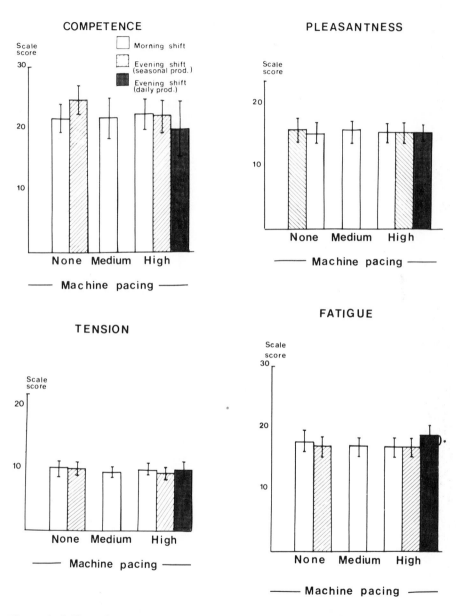

Figure 4. Self-rated competence, pleasantness, tension and fatigue at the end of the work shift by workers at various levels of machine pacing.

We can also ask whether the restricted work content, with a low level of variation and high repetitiveness in all the groups, is so dominating that it hides the possible effect of machine pacing. This hypothesis may be partly adequate, even though recent theoretical frameworks underline both repetitiveness and the lack of self-determination as major stressors at work (Karasek 1979).

Some uncontrolled background factors could be one reason for the present findings. The workers are paid in part on a piece-work basis. The exact amounts of their wages could not be controlled in the study. This type of wage system may lead to an acceleration of workspeed also in the less machine-paced groups.

A common problem in field studies, especially cross-sectional ones, is selection to and from occupations. The work experience of all the study groups was relatively long, which may have allowed sufficient time for selection and adaptation.

A final possible explanation is the type of machine lines and work arrangement in the printing industry. The lines are generally short: only rarely do they comprise more than six workers. The positions of the workers are often somewhat flexible, thus allowing discussion. At certain positions of the lines the workers may momentarily replace one another. This flexibility may partly explain the findings.

References

Cox, T., 1980, Repetitive work. In *Current Concerns in Occupational Stress*, edited by C. Cooper and R. Payne (Chichester: John Wiley), pp. 23–41.

Johansson, G., Aronsson, G., and Lindström, B. O., 1978, Social, physiological, and neuroendocrine stress reactions in highly mechanized work. *Ergonomics,* **21,** 583–599.

Kalimo, R., Leppänen, A., Seppälä, P., Louhevaara, V., and Koskinen, P., 1981, *Work organization, production technology and mental strain* (in Finnish) Reports of the Institute of Occupational Health, Helsinki.

Karasek, R. A., 1979, Job demands, job decision latitude, and mental strain: Implications from job redesign. *Administrative Science Quarterly,* **24,** 285–308.

Mackay, C., Cox, T., Watts, C., Thirlaway, M., and Lazzerini, A. J., 1979, Physiological correlates of repetitive work. In *Responses to Stress, Occupational Aspects*, edited by C. Mackay and T. Cox (Guildford: IPC Science and Technology Press), pp. 129–141.

Subjective reports of stress and strain as concomitants of work

By R. G. Pearson

North Carolina State University, Departments of Industrial
Engineering and Psychology, Raleigh, North Carolina 27650, USA

Psychometric scales which have been used to assess work-related stress and strain are reviewed. Subjective reports of strain can be contrasted with other states which are often used synonymously, and/or confused, with strain, e.g. fatigue, annoyance, discomfort, boredom and anxiety. The relation of subjective states to work, and especially paced tasks, is discussed. The paper addresses work situations where the question can be raised whether stress is all that bad. Like 'fatigue', often 'strain' is a natural by-product of work, one without health consequences and bearing no relationship to work decrement. Indeed, in some situations, the amelioration of strain complaints can be achieved through work, i.e., energy expenditure.

1. Introduction

From the viewpoint of an ergonomist, this author's concern with the topic of stress relates to its effect on the performance, safety, health, and wellbeing of the worker—a concern obviously shared by the organizers of this conference and its USA sponsor, NIOSH. With regard to the goals of the US Occupational Safety and Health Act of 1970, the author has for many years equated (1) the phrase involving protection of the worker at his workplace from hazards to 'wellbeing' with (2) a concern for the mitigation of stress and those related symptoms of strain.

The principal purpose of this paper is to review the subjective tools used to assess work-related stress and strain. The focus is intended to be upon psychometrically based scales (e.g. those developed using Likert, Thurstone, or Guttman methodologies) and upon dimensions identified by such techniques as factor analysis; as such, treatment of items contained in stress or job satisfaction surveys will be de-emphasized. While a number of scales exist which measure a variety of strain-related dimensions, no single scale of subjective strain exists which can be used at the workplace to assess a worker's fitness to perform or to continue at his task. Thus, the reader may anticipate that the author's concluding remarks will reflect both disappointment and challenge.

One of our problems, of course, is the elusive nature of the stress concept.

Subjective measures of strain, depending on the research focus, range over a variety of dimensions, e.g. fatigue, effort, discomfort, tension, anxiety, boredom, annoyance, etc. Perhaps, then, it is well that we view strain as having a multidimensional character. Stating this another way, indices of strain must be related to the types or sources of stress encountered at the workplace. In this regard we must examine the task itself, the workplace and its environment, the worker, and both intra-organizational and extra-organizational factors in order to 'tailor' our strain measures to specific stresses, as these exist. The studies to be reviewed next, accordingly, reflect a variety of stress factors and dimensions of strain.

2. Studies of stress and subjective strain

The 'Social Readjustment Rating Scale' (Holmes and Rahe 1967) was developed on the basis of clinical evaluation of major life changes of over 5000 patients experiencing 'stress-related' illness. Weights are assigned to 'life events' such as death of spouse, divorce, illness, being jailed, loss of job, financial problems, changes in work responsibilities, etc.—43 items in total. Events are to be reported for the period of the previous year. Scores are taken to be an indicator of the degree of strain a person is experiencing at the time of testing, higher scores being associated with an increased probability of contracting illness. As a source of caution and guidance the scale appears useful, but its predictive utility for industrial use is unproven. Less than 25% of the items are job-related. Additionally, the scale does not take into account an individual's ability to cope with stress.

Cox and Mackay (1979) conducted a study of the stress associated with repetitive work involving over 500 shop-floor workers. A job description checklist involving 55 adjectives was used to assess workers' perception of their work. Factor analysis of the survey data identified four major dimensions labelled as (1) pleasantness, (2) tedium, (3) pressure, and (4) difficulty. These four factors provide some insight into the types of strain associated with the stress of repetitive work as perceived by workers.

Perhaps the most comprehensive study to date of occupational stress and strain is that conducted by Caplan and associates and reported in the NIOSH publication *Job Demands and Worker Health* (1975). This cross-sectional study involved 23 jobs, a sample of 2010 men, and a questionnaire which assessed (1) 20 job stresses, (2) 17 strains, and (3) a number of demographic and personality variables. Additionally, physiological measures were taken on 390 of the workers for eight of the 23 jobs. Results supported a strong relationship between job stresses and subjective measures of strain. However, only moderate relationships were found between the strain findings and indices of illness, e.g. cardiovascular disease and ulcers.

This study by Caplan and associates utilized seven questionnaire measures of psychological strain. The individual scales, and their number of items, were as follows: job dissatisfaction (7); workload dissatisfaction (3); boredom (3); somatic complaints (10); depression (6); anxiety (4); and irritation (3). Estimates for reliability for all scales were quite respectable, ranging from $0 \cdot 75$ to $0 \cdot 86$. Of particular interest are the data on 79 men involved in machine-paced assembly work. As contrasted with the total sample, the psychological strain measures for this group reflected higher amounts of job dissatisfaction, boredom, workload dissatisfaction, somatic complaints, anxiety, and depression. Such workers also showed an above-average frequency of recent visits to their industrial dispensary. Stress factors which were higher for this group, and which can be related to the strain responses, included: under-utilization of abilities; job-future ambiguity (related to concerns for job security and skill utilization in the future); responsibility for persons—poor fit (a measure of discrepancy between the amount of actual job responsibility for the wellbeing of others and that amount desired by the worker); and job complexity—poor fit (the discrepancy between the worker's actual job complexity and his preference for complex activity).

With regard to the strains of machine-paced work, Caplan and associates conclude that workers '. . . on the machine-paced assembly lines are the most bored and the most dissatisfied with their work load.' Also among these workers, complaints of anxiety, depression, irritation, and somatic complaints were the highest for all occupations involved in the study. A major contribution of this study is the insight which it provides concerning those causative factors related to strain responses. Through examination of these causative factors the opportunity for exploring preventive courses of action presents itself. Overall, the study makes a valuable contribution to our understanding of the relationships among (1) occupational stress, (2) psychological and physiological strain responses, and (3) stress-related illnesses. Additionally, while not discussed in detail here, the study points out the contribution, or relationship, of demographic and personality variables to strain responses. In this context the importance of considering individual differences in strain responses must be emphasized. We must take note that subjective strain responses involve a perceptual process—one must perceive a situation as stressful in the first place; if not, then stress as such does not exist for that person. In short, a stressful situation may lead to stress responses in one person, but not in another.

With regard to the issue of individual differences, a study in our laboratory involving 166 subjects examined the annoyance effects of a variety of stressful noise exposures (Pearson *et al.* 1974). Subjects were tested in small groups while seated in an experimental 'living room' environment. Using a 25-point Annoyance Rating Scale shown in figure 1 we found, as expected, different mean annoyance ratings for the six noise stimuli. Across noise stimuli, the mean rating was $11 \cdot 76$; means for the individual noises ranged from a low of $7 \cdot 96$ (propeller aircraft) to a high of $19 \cdot 6$ (pneumatic chipping hammer) even though the peak sound-pressure level was equated for all noise stimuli at 82 dB. However, the greatest amount of variance in annoyance ratings was attributed to the subjects themselves. Mean individual ratings (across the six stimuli) ranged from 1·65 to a high of 22·20 on the 25-point scale. In brief, some subjects found the noises extremely annoying, while others found them hardly objectionable.

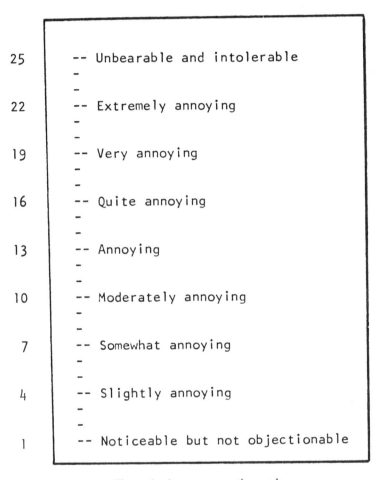

Figure 1. Annoyance rating scale.

In a related laboratory study (Cohen *et al.* 1974) we examined the relationships among workpace, noise stress, and subjective responses. Six subjects performed an information-processing task. A random sequence of four digits which changes on each trial is presented via a digital display, and is then followed by illumination of one of four vertically arranged stimulus lights. The subject is required to make a code translation (i.e. recall the code and then relate this to the light position) and then react by pressing one of four response buttons. All subjects performed the task for one hour under all combinations of three workpace conditions (30, 40, and 50 trials/min) and two noise conditions (noise; quiet), i.e., under six different experimental conditions. The noise stress (presented to the subject through earphones) involved rapid, intermittent pulses of constant duration, broadband 100 dB(A) noise separated by rapid, variable internoise intervals. Such noise was subjectively similar to that associated with some office and computing equipment environments as well as with certain automated or semi-automated assembly processes found in industry.

Figure 2 shows the highly significant interaction effects between noise stress and workpace. In short, under conditions of noise, the performance cost increases disproportionately as a function of faster workpace. Generalizing to the industrial task, we all know the importance of determining an optimal workpace. If it's too slow, we face the problem of boredom; if it's too fast, then we refer to the situation as one involving 'speed stress'.

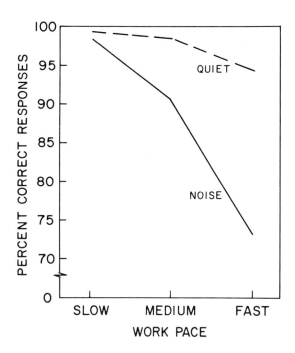

Figure 2. Interaction effects of noise and workpace on task performance.

Subjective comments were also obtained from the subjects at the conclusion of the study. In particular, the added noise stress made it difficult for subjects to attend to their task and to sustain their concentration over the work period.

Also, the fast workpace itself was perceived as stressful, and the added noise increased the feelings of strain. Finally, the subjects reported a number of physiological reactions indicative of a high state of autonomic arousal (or stress), e.g., profuse palmar sweating, neck and shoulder muscle tension, hand and finger cramps, blanching of the hand and fingers, and feelings of finger coolness and numbness. Even though the task was of short duration, was relatively simple, and involved only a small amount of energy expenditure, it is interesting to note the extent to which physiologically based strain responses can be generated by the combined stress of a fast workpace and noise. Clearly, there are implications here for the health and wellbeing of workers exposed to such conditions.

3. Retrospect and prospect

During the mid-1950's, the author worked in a laboratory component of the USAF School of Aviation Medicine, where one of our principal research concerns was the problem of aircrew fatigue. More specifically, our focus was on the work decrement observed over time, which characterizes so many tasks involving perceptual-motor skill and monitoring behaviour, i.e., sustained attention. The task most commonly used in this research was the USAF-SAM Multi-Dimensional Pursuit Task. This apparatus involved the concurrent monitoring of four meter displays and co-ordinated operation of four associated controls. Two equivalent forms of a 'Feeling Tone Checklist' were developed (Pearson and Byars 1956) to assess subjective fatigue and were used to assess the relationship between subjective state and performance; Form A of this checklist is shown in figure 3. For 100 experimental subjects (basic trainees) who performed the task for three hours (following learning to asymptote and a rest break) the average subjective fatigue scores declined over time, significantly so in comparison to a control group of 100 subjects who remained 'on alert' in a lounge area adjacent to our test room but who did not engage in task activity (Pearson 1957). While it was apparent that there was some parallelism between the downward course of task performance and the feeling tone data, one must recall we were dealing with averages. Indeed, there were great individual differences in both performance and subjective report. Consequently, when correlations were computed to examine relationships between (1) the current level of a subject's performance or (2) the course of his future performance, the resulting values were found to be not significantly different from zero. This author concluded then that the way a subject says he feels prior to the three-hour task and the way he performs the task are not necessarily related, nor do the subject's feelings necessarily parallel his performance.

Later, in an unpublished study for the Military Air Transport Service, the author assessed pilot performance and subjective fatigue during 14- to 16- hour

FEELING   TONE   CHECKLIST,   FORM   A

NO.	BETTER THAN	SAME AS	WORSE THAN	STATEMENT
1	( )	( )	( )	SLIGHTLY TIRED
2	( )	( )	( )	LIKE I'M BURSTING WITH ENERGY
3	( )	( )	( )	EXTREMELY TIRED
4	( )	( )	( )	QUITE FRESH
5	( )	( )	( )	SLIGHTLY POOPED
6	( )	( )	( )	EXTREMELY PEPPY
7	( )	( )	( )	SOMEWHAT FRESH
8	( )	( )	( )	PETERED OUT
9	( )	( )	( )	VERY REFRESHED
10	( )	( )	( )	READY TO DROP
11	( )	( )	( )	FAIRLY WELL POOPED
12	( )	( )	( )	VERY LIVELY
13	( )	( )	( )	VERY TIRED

Figure 3. Checklist format.

flights between the USA East coast and air bases in Europe. Again, there were no apparent indications of relationships between subjective state and indices of flying proficiency. Since that time, Bryce O. Hartman and colleagues at the USAF School of Aerospace Medicine have continued to use the checklist forms in studies of aircrew stress. Briefly, these studies reveal the same parallelism between subjective state and time on task *as averaged across subjects*. Only when the task duration extends to 22 to 24 hours or more do we see a close correspondence for individual subjects—that is, the subject reports he is 'dead tired' and his performance has become disorganized in the fashion described by Sir Frederick Charles Bartlett as 'skill fatigue' (Bartlett 1943).

There is a moral to this story which can be generalized to the subject of stress. S. Howard Bartley, many years ago, in his classic text with Eloise Chute (Bartley and Chute 1947), equated subjective fatigue with 'task aversion'. Briefly, feelings of fatigue are a natural concomitant of work. The longer we work, the more we feel task-aversive and desire to do something else. But this doesn't mean we can't perform our task! Indeed, for many of us we turn to other work (in the sense of energy expenditure) at the end of our day, e.g. gardening, jogging, bowling and the like. In short, neither fatigue nor stress is necessarily bad. And the cure for such feelings of lethargy or strain, interestingly enough, for those engaged in sedentary occupations is *work*.

For many years since the author's early experiences in the Air Force, he has lectured on the need for what has been termed a 'Handy-Dandy Fatigometer'. This would be some sort of electrode device attached to the body with an associated meter display that would indicate when an operator is 'fatigued' to the point that he should be removed from his task. Perhaps a similar 'Stresso-meter' could be developed. Clearly, today we do not have either of these devices. The variety of fatigue scales available do not permit such individual prediction of unsuitability to work in the typical industrial situation and no specific checklist of strain symptoms exists for similar purposes. It would be nice if we had such a device, something similar to Nigel Corlett's approach (Corlett and Bishop 1976) to assessing postural discomfort, that would permit us to assess those levels of strain which would indicate when the worker should be removed from his task relative to concerns for lost productivity and costs to health.

In summary, strain complaints, whether of job dissatisfaction, fatigue, tension, anxiety, depression, or annoyance should be regarded as 'epipheno-mena' (to use Bartley's term)—that is, taken merely as indicative of signs of potential stress factors at the workplace. As such they must be examined and qualified in terms of a host of other moderating and interacting factors. Not-withstanding, we should also guard against taking all subjective strain responses too seriously lest we forget that many complaints are a natural concomitant of working and living—and are not necessarily of consequence to health.

References

BARTLETT, F. C., 1943, Fatigue following highly skilled work. *Proceedings of the Royal Society,* B, **131,** 247–254.

BARTLEY, S. H., and CHUTE, E., 1947, *Fatigue and impairment in man* (New York: McGraw-Hill).

CAPLAN, R. D., COBB, S., FRENCH, J. R. P., Jr, VAN HARRISON, R., and PINNEAU, S. R., Jr, 1975, *Job Demands and Worker Health* (Washington, D.C.: National Institute for Occupational Safety and Health, U.S. Department of Health, Education, and Welfare, Publication 75-160).

COHEN, H. H., CONRAD, D. W., O'BRIEN, J. F., and PEARSON, R. G., 1974, *Effects of noise upon human information processing* (Hampton, Virginia: NASA CR 132469).

CORLETT, E. N., and BISHOP, R. P., 1976, A technique for assessing postural discomfort. *Ergonomics,* **19,** 175–182.

COX, T., and MACKAY, C. J., 1979, The impact of repetitive work. In *Satisfactions in Work Design,* edited by R. G. Sell and P. Shipley (London: Taylor & Francis), pp. 101–112.

HOLMES, T. H., and RAHE, R. H., 1967, The social readjustment rating scale. *Journal of Psychosomatic Research,* **11,** 213–218.

PEARSON, R. G., 1957, *Task proficiency and feelings of fatigue* (Randolph Field, Texas: School of Aviation Medicine, USAF, Report 57-77).

PEARSON, R. G., and BYARS, G. E., Jr, 1956, *The development and validation of a checklist for measuring subjective fatigue* (Randolph Field, Texas: School of Aviation Medicine, USAF, Report 56-115).

PEARSON, R. G., HART, F. D., and O'BRIEN, J. F., 1974, *Individual differences in human annoyance response to noise* (Hampton, Virginia: NASA-CR 14491).

Use of proportion-of-baseline measures in stress research

By A. C. Bittner, Jr

Naval Biodynamics Laboratory,
P.O. Box 29407, Michoud Station,
New Orleans, Louisiana 70189, USA

The purpose of this report was to provide a preliminary basis for evaluating the use of percentage-of-baseline and related measures in stress research. Possible fine-structure responses of individual subjects, including proportional responses, were delineated and methods for their identification described. Propagation of error with proportion-of-baselines was outlined and compared to recent research. It was concluded that the results contraindicate the arbitrary use of proportion-of-baseline scores for stress research.

1. Introduction

The last decade has seen a growing body of studies employing proportion-of-baseline measures in human-stress research (e.g., Alluisi *et al.* 1977, Collins 1973, Morgan *et al.* 1980, Morrissey and Bittner 1975). Accompanying these growing applications has been evidence questioning the use of proportion-of-baseline measures. Examination of a portion of such evidence and selected implications is the focus of this report.

Empirical evidence questioning the use of proportion-of-baseline measures has been found in integrative studies of heat and vibration effects (Morrissey and Bittner 1975, Ramsey and Morrissey 1977). In the most pertinent study, Ramsey and Morrissey attempted to use decremented proportions in their investigation of heat exposure on mental performance. Following the lead of Collins (1973), they defined the proportional decrement Z_{ki} as

$$Z_{ki} = (X_{ki} - X_{0i})/X_{0i} \tag{1}$$

where X_{0i} and X_{ki} were performance scores of the *i*-th unit (mean subject) for respective baseline (0) and stress $(k \rangle 0)$ conditions. Ramsey and Morrissey could have used the linearly related 'percentage decrement' $(100Z_{ki})$ or 'proportion scores' (X_{ki}/X_{0i}) but they rightly noted that results would be isomorphic. Once proportional decrements had been computed over the body of reports, the researchers cast the results into a series of Wet Bulb Global Temperature (WBGT) by Exposure Time graphs. Graphs were divided by tasks or task types with one each for different aspects of 'mental performance'. The result of the graphical analyses of proportional decrements was chaos which initially defied interpretation.

Because of potential implications for future efforts, an attempt at determining causes of the chaos was initiated by Morrissey and this author. Results of this analysis revealed several partially interwoven sources. The largest source appeared to be due to systematic error structures induced (propagated) through the decremental equation (1). Indeed, in examining raw data collected by Ramsey, large variations in proportional decrements could be seen where large relative errors existed in underlying subject data. Co-variation of decrements also was apparent where common baselines had been employed to calculate decrements. These findings had precedence with Morrissey and Bittner (1975) having noted a suggestion for such a structure in their earlier vibration investigation. The contrast in magnitudes, however, led to the conjecture of fundamental differences in the fine-structure responses of individual subjects under the two environments. Overall, the investigation of heat-stress decrements data focused attention on both fine-structure responses and propagation of errors.

The goals of this report are (1) to delineate possible fine-structure responses to stress and methods for their identification, and (2) to study analytically the propagation of errors with proportion-of-baseline measures. The purpose of this report is to provide a preliminary basis for evaluating the use of proportion-of-baseline and other measures in stress research.

2. Fine-structure responses to stress

The approach used in this section initially will be to outline a family of 'fine-structure responses' and subsequently illustrate their consideration with data from a vibration experiment. Table 1 shows linear and proportional forms of three models of individual (*i*-th) subject response to stress. Examining the leftmost column of this table, three classes of response can be seen, with Model I reflecting a constant additive effect, Model II a proportional effect, and Model III combined effects. These models, while believed descriptive of stress data, were suggested respectively by the usual model for analysis-of-variance, the proportion-of-baseline measure, and the model for linear cor-relation. The middle column of this table shows proportional reformulations of the three models. Lastly, the rightmost column shows the relationships between measurement errors for the respectively linear and proportional forms of particular models. Overall, table 1 outlines three competitors for characterizing subject response with error relationships distinguishing linear and proportional forms.

Considering the models and forms of table 1, little theoretical rationale can be seen for *a priori* choice of any model. The choice of linear versus

Table 1. Linear and proportional formulations of three models relating baseline (X_0) and stress (X_1) observations.

Models	Linear form	Proportional form	Relationships between (ϵ, ϵ') errors
I	$X_{1i} = \alpha + X_{0i} + \epsilon_i$	$X_{1i}/(\alpha + X_{0i}) = 1 + \epsilon'_i$	$\epsilon_i = (\alpha + X_{0i})\epsilon'_i$
II	$X_{1i} = \beta\, X_{0i} + \epsilon_i$	$X_{1i}/(X_{0i}) = \beta + \epsilon'_i$	$\epsilon_i = (X_{0i})\epsilon'_i$
III	$X_{1i} = \alpha + \beta X_{0i} + \epsilon_i$	$X_{1i}/(\alpha + \beta X_{0i}) = 1 + \epsilon'_i$	$\epsilon_i = (\alpha + \beta X_{0i})\epsilon'_i$

proportional form also appears to be equivocal, although the linear forms are easier to 'fit' to a set of subjects' data. However, this choice must not be arbitrarily made because an improper model results in inflation of estimated error with model lack-of-fit variation. Hence, more subjects than necessary may be exposed to a hazardous environment to obtain a given level of apparent precision. Similar arguments can be advanced regarding the choice of form (weighting of error). Obviously, the choice of model and form requires an empirical basis.

The choice of model and form can be made by either formal statistical analysis, graphical analysis, or a combination of analyses. Cochran (1977) has suggested the use of a combination of approaches for ratio estimators (Model II). To illustrate these approaches, unpublished data provided by J. C. Guignard will be utilized. Figure 1 shows the performances of 14 volunteer personnel which were obtained under baseline and vibration (8 Hz, 0·22 Grms) conditions on three tasks. Coding represented a cognitive task, tapping a gross motor task and aiming a fine motor task (Carter *et al.* 1980). Prior to data collection, volunteers were given five practice trials on the three tasks during the day preceding the study and one warm-up trial on the day of the study. Data were then taken in a pre (baseline) condition, a per (vibration) condition, and post (baseline) condition. The average of the (*i*-th subject's) pre and post scores for each task made up the baseline scores (X_{0i}) with the per score the corresponding (X_{1i}) vibration value. For illustrative purposes, other details of the experiment may be ignored.

Figure 1 provides the necessary data for graphical analysis. Examining this figure, it can be seen that performance decrements are apparent for all measures. Tapping, where larger scores reflect lessened performance, appears to be clustered parallel to the baseline-vibration isoperformance line with homogeneous dispersion along its length. The parallel array suggests Model I and the homogeneous dispersion of the cluster along its length indicates the linear form. Models II and III, it is pertinent to note, would be suggested by clusters which were not parallel to the isoperformance line, with Model II

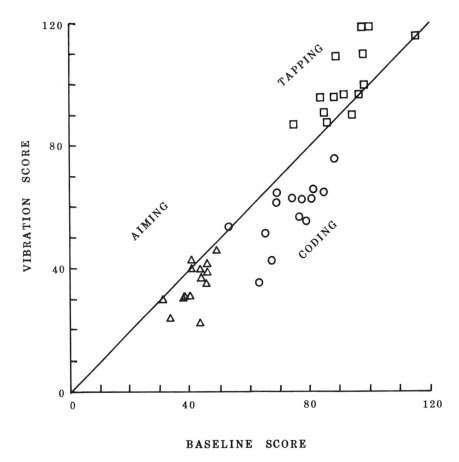

Figure 1. Baseline (X_0) versus Vibration (X_1) performance on aiming (number $\times 10^{-1}$), coding (number) and tapping (10^2 s/correct) tasks.

indicated by a cluster which projects towards the origin. In addition, proportional forms would be suggested by heterogeneous dispersions which broaden with increasing values. Examining coding, Model II is suggested as its cluster projects toward the origin; the linear form appears preferable as increasing dispersion for lower values contraindicates a proportional form. The aiming cluster also shows a tendency toward heterogeneity for lower values, contraindicating a proportional form. Taking into account this dispersion, aiming appears to be best represented by Model I. Altogether, graphical analyses, while not definitive, have provided some insight into the fine structure of subject responses.

Statistical analyses, analogous to the graphical analyses, can be used to identify model and form. In particular, linear least-squares analysis can be used to fit the linear forms of Models I, II and III; and non-linear least-squares can be used with proportional forms. Residual errors, from any fitted model, can be examined for distribution and homogeneity by a number of methods. Applicable methods include calculation of residual moments and correlation of absolute residual errors with stress scores. Such methods were applied to the tapping, coding, and aiming data shown in figure 1. Linear forms were indicated for all three data sets, and the results of the graphical analysis were supported. In particular, tapping was found to be best represented by Model I, $X_{1i} = 0 \cdot 79 + X_{0i} + \epsilon_i$, with a residual variance of $0 \cdot 642$ vice $0 \cdot 671$ and $0 \cdot 674$ respectively for Models II and III. Coding was found best represented by Model II, $X_{1i} = 0 \cdot 79 X_{0i} + \epsilon_i$, with a residual variance of $54 \cdot 8$ vice $58 \cdot 9$ and $61 \cdot 1$ respectively for Models I and III. Lastly, aiming was found best represented by Model I, $X_{1i} = X_{0i} - 62 \cdot 1 + \epsilon_i$, with a residual variance of 1383 vice 1539 and 1444 respectively for Models II and III. Obviously, statistical fitting of model and form provides either an adjunct or alternative to graphical analysis. Statistical analysis is preferable to a graphical analysis because its results are less subjective.

3. Propagation of errors with proportion-of-baselines

The approach used in this section will be to provide formulations of variances and co-variances propagated with proportion-of-baseline measures and to consider implications of these formulations. Formulation of propagated errors requires the use of a model for derivation. Following Bittner (1980), let (X_{0i}, X_{1i}, X_{2i}) be a set of independent observations for the i-th subject from a sample of N. (The value X_{0i} will be defined as a baseline measure with X_{1i} and X_{2i} defined as stress conditions measures following earlier notation. In addition, assume that the expected means and variances for all subjects are equal to group values. Lastly, assume equal stress means ($\mu_1 = \mu_2$) and variances ($\sigma_1^2 = \sigma_2^2$) and define the ratios of stress to baseline parameters by $p = \mu_k / \mu_0$ and $q = \sigma_k^2 / \sigma_0^2$ ($k = 1,2$). Then from a 'total differential' approach (Deming 1943), it follows that the large sample ($N \to \infty$) mutual expected variance of the proportion scores $Z_{1i} = X_{1i} / X_{0i}$ and $Z_{2i} = X_{2i} / X_{0i}$ is

$$\text{Var}(Z) = (p^2 + q)\sigma_0^2 / \mu_0^2 \tag{2}$$

and their co-variance is

$$\text{Cov}(Z_1, Z_2) = p^2 \sigma_0^2 / \mu_0^2 \tag{3}$$

where p and q are as defined above (Bittner 1980). Equations (2) and (3), pertinently, are analogous for proportional decrement and percentage scores. Examining equations (2) and (3), the systematic error structures seen in study of the Ramsey and Morrissey (1977) data can be explained. The variance (2), in particular, can be seen to be proportional to the relative error, supporting the earlier results. The proportional component (p^2) of this expression attests to the complexities when stress means and variances vary from baseline. In addition, the co-variance expression (3) sheds light on the earlier finding of correlations between proportion scores calculated from the same baselines. Altogether, the variance and co-variance structures propagated with proportion-of-baseline scores appear complex and potentially misleading.

An example of the misleading character of proportion-of-baselines is provided in the recent work of Morgan *et al.* (1980). In their study, they reported the reliability of changes in percentage-of-baseline performances over four continuous work and sleep-loss exposures. On 13 tasks, Morgan *et al.* obtained Kendall's coefficients of concordance ranging from $W = 0 \cdot 29$ to $0 \cdot 85$ with an average of $W = 0 \cdot 68$. These concordances *appeared* to be significant ($p \langle 0 \cdot 05$) in all but one case and were taken to imply consistent individual decrements across conditions. However, the propagated error equations (2) and (3) indicate an expected correlation between proportional scores based with the same baseline score. The correlation between two such proportional scores, calculated from the ratio of equation (3) and (2), is

$$r_z = p^2/(q + p^2) \tag{4}$$

where p and q are as defined above (Bittner 1980). Letting $p = q = 1$ in equation (4), an expected correlation of $r_z = 0 \cdot 5$ is expected. This value of r_z also can be arrived at by a detailed study of Morgan *et al.*'s results; it can also be shown to imply an average concordance of $W = 0 \cdot 63$ (Bittner 1980). Compared with the average over the 13 tasks reported by Morgan *et al.* of $W = 0 \cdot 68$, little difference is seen. Hence, the Morgan *et al.* (1980) results appear to be largely an artefact of the proportion-of-baseline measure. Obviously, the propagated error structures for proportion-of-baselines must be carefully considered prior to the use of such measures.

4. Discussion

The results of the previous sections lead to two conclusions. Firstly, fine structures should guide choice of model and model form in stress research. Secondly, the propagation of errors with proportion-of-baseline measures should be considered prior to their use. Altogether, the results contraindicate the arbitrary use of proportion-of-baseline scores for stress research.

This work was conducted under Navy Work Unit MF58.524-026-5027. The opinions expressed are those of the author and do not necessarily reflect those of the Department of the Navy.

References

ALLUISI, E. A., COATES, G. D., and MORGAN, B. B., Jr, 1977, Effects of temporal stressors on vigilance and information processing. In *Vigilance: Theory, Operational Performance and Physiological Correlates*, edited by R. R. Mackie (New York: Plenum Press), pp. 361–424.

BITTNER, A. C., Jr, 1980, *Artificially induced individual differences with percentage-of-baseline measures,* New Orleans, LA, Naval Biodynamics Laboratory, unpublished manuscript.

CARTER, R. C., KENNEDY, R. S., and BITTNER, A. C., Jr, 1980, Selection of performance evaluation tests for environmental research. *Proceedings of the 24th Annual Meeting of the Human Factors Society*, Los Angeles, pp. 340–343.

COCHRAN, W. G., 1977, *Sampling Techniques*, 3rd ed. (New York: John Wiley), pp. 158–160.

COLLINS, A. M., 1973, Decrements in tracking and visual performance during vibration. *Human Factors*, **15**, 379–398.

DEMING, W. E., 1943, *Statistical Adjustment of Data* (New York: Dover Publications), pp. 37–48.

MORGAN, B. B., WINNE, P. S., and DUGAN, J., 1980, The range and consistency of individual differences in continuous work. *Human Factors*, **22**, 331–340.

MORRISSEY, S. J., and BITTNER, A. C., Jr, 1975, *Effects of vibration on humans: Performance decrements and limits*, TP-75-47, Point Mugu CA, Pacific Missile Test Center.

RAMSEY, J. D., and MORRISSEY, S. J., 1977, A composite view of task performance in hot environments. *Proceedings of the 21st Annual Meeting of the Human Factors Society*, pp. 73–77.

Section 5.   Select issues in machine-paced research

A taxonomic framework for the description and evaluation of paced work

By M. J. DAINOFF†, J. J. HURRELL, Jr† and A. HAPP‡

†Applied Psychology and Ergonomics Branch,
Division of Biomedical and Behavioral Science,
National Institute for Occupational Safety and Health,
Centers for Disease Control,
Public Health Service,
Department of Health and Human Services,
Cincinnati, Ohio, USA

‡Department of Psychology,
Miami University,
Oxford, Ohio, USA

It is argued that a standardized system of description and classification for different kinds of machine-pacing task is required if progress is to be made in relating attributes of such tasks to occupational stress symptoms. It is proposed that a classification of tasks with respect to operator control of work initiation and duration would provide a comprehensive and scientifically useful framework for further research.

1. Introduction

Attempts to understand the relationship between attributes of machine-paced work and symptoms of occupational stress have been hampered by a general lack of consistency among authors regarding definitions and descriptions of task structures (see Hurrell 1981 for a review of this material). Moreover, it is likely that this lack of consistency may be responsible for at least some of the contradictions and confusions which appear in the existing literature on this topic.

It would therefore seem useful to propose a relatively standardized terminology for this subject area so that researchers could communicate with one another in as consistent a fashion as possible, while at the same time providing a uniformly complete physical description of those aspects of the work which are particularly pertinent to issues of occupational stress. Accordingly, the following terminological review and proposed taxonomy of paced-work is presented.

2. Basic terminology

There does not seem to be much difficulty in accepting, as a fundamental definition of paced work, the concept of work in which the worker is required to respond at a rate other than that which would be self-selected. Thus, for example, for Franks (1974) a task will be considered paced if '... there exists external sensory stimuli in the form of temporal signals of any nature which do not depend on a reaction for their presentation'. It will be noted that this definition implies a formal separation of *machine action* from *operator action*; while at the same time specifying that the interaction between those temporal factors which determine the former and latter define the nature of the task (Happ 1981, Conway *et al.* 1977). Thus, in a completely machine-paced task, rate of presentation is under machine control and the action of the operator in no way influences the presentation of succeeding parts (or information). In contrast, in the completely unpaced task, the rate of work depends entirely on the action of the operator.

The temporal parameters determining both operator and machine function can, in general, be specified more completely. With respect to description of operator function, we can define *operator cycle time* (*OC* in figure 1) as the time elapsed between commencement of work on successive parts or information sources (Sury 1967). This time increment can, in turn, be divided into two further components; *service time* (*S*) and *operator delay time* (*OD*) (Sury 1967). Service time refers to that portion of the operator cycle time during which the operator is actually working on a part or processing information. Operator delay time, on the other hand, refers to that portion of the cycle during which the operator must wait for a new part or information source.

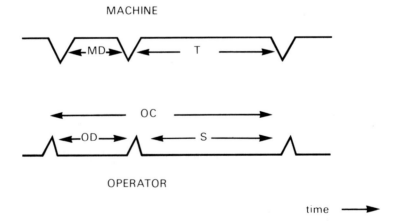

Figure 1. Temporal parameters in paced work. *OC* = operator cycle time, *S* = service time, *OD* = operator delay time, *T* = tolerance time, *MD* = machine delay time.

With respect to machine function, *tolerance time* (*T*) refers to the length of time a part or information source is available for processing by the operator (Sury 1967, Franks 1974). Finally, *machine delay time* (*MD*) refers to that part of the machine cycle in which the part/information is not available for processing.

Each of the parameters indicated in figure 1 is a variable which may function in either a deterministic or stochastic fashion, depending on the particular work.

When work is completely paced, tolerance time is likely to be relatively fixed, whereas service times will have some statistical distribution (normal or skewed) with a variance which is non-trivial. When work is unpaced, neither variable is likely to be deterministic except in the sense that the offset of machine tolerance time will typically be caused by the self-termination of operator service time.

3. Varieties of pacing

Given the above parameters, a number of different types of pacing are possible, and several distinctions in classification have been made in the literature. Conrad (1954) differentiated between 'rigid systems' and 'systems with margins'. Dudley (1962) described the distinction between these types of pacing as follows:

> In an extreme case, the operative may be rigidly paced by a machine, in that the time allowed to perform the operation is equal to the time required for its completion and every article or component must be dealt with. In other cases, a few misses may be allowed, a few faults tolerated, or a little waiting time introduced. Alternatively, the work itself may be allowed to accumulate at the work station to permit some degree of flexibility on the part of the operative.

Murrell (1963) described two slightly different types of pacing which he referred to as Type 1 and 2 pacing. In Type 1 pacing, there is a time period in which some operation must be carried out; but little or no work is performed by the operative while the machine is indexing or processing the part. This type typically occurs in situations where workers perform tasks which support machine functioning (i.e., feeding machines, removing processed part). In Type 2 pacing, work is performed during the period that the machine indexes and must be completed at the time that the machine is ready to be fed (i.e., work is performed in synchrony with the machine cycles). This type occurs when operatives must remove a part from a belt, process it and return it to the belt before the next part passes out of reach.

Different investigators have described other variants of machine pacing. Buxley *et al.* (1973), for example, identified three types of flow line where pacing may exist: (1) single-model lines (where only one model or type of product is produced); (2) multi-model lines on which two or more similar types of model or products are processed separately in batches; and (3) mixed-model lines where two or more similar models or products are produced simultaneously. Conway *et al.* (1977) further describe a variation which combines elements of both unpaced and paced work. In this process, cycles are initiated by the operatives. However, once the cycle is started, the worker is paced through a rapid sequence of motions. Finally, Rohmert and Luczak (1973) have extended the concept of pacing to include information-processing tasks. In such 'paced-information tasks', the service-time component can be partitioned into an information/decision component, and a motor component.

From the above discussion, it should be obvious that numerous varieties of pacing exist and that different paced systems may require vastly different amounts of cognitive and motor activity from the worker. Unfortunately, many of the studies concerned with the effects of pacing fail to adequately document the specific characteristics of the system being examined. Thus, one often knows little about the independent variable being considered except that it is something called pacing. Further, the distinctions made among different kinds of pacing have not been demonstrated to be either comprehensive enough or to be theoretically useful as explanatory concepts in understanding the potentially stressful aspects of paced work. The following system is proposed in an attempt to remedy both of the above deficiencies.

4. Proposed taxonomic system

The extent to which the worker has control over aspects of his daily work environment will form the basis of the classificatory system to be proposed. Karasek (1979), in his own classification of general work situations, has shown that specification of the extent of such control (or job decision latitude, in his terms) can, along with knowledge of level of job demands, be very effective in predicting the magnitude of occupational stress symptomatology. In particular, he showed that situations with high demand and low decision latitude (control) result in high levels of symptoms.

In the case of paced work, operator control is manifested along two orthogonal dimensions: control over the *initialization* of the work cycle, and control over the *duration* of the work cycle. The extent to which the worker, as opposed to the machine, has control over either or both of these functions is reflected in the four-quadrant classification scheme indicated in figure 2.

In Quadrant I (QI), tasks are initiated by the machine, but work time is under the control of the operator. Such a task could be a telephone switchboard operation in which calls arrive under the control of the machine (external environment), but the operator determines how long it takes to process a call. Work of this type has not generally been studied in the pacing literature up to this point, but recent research on secretarial/clerical workers

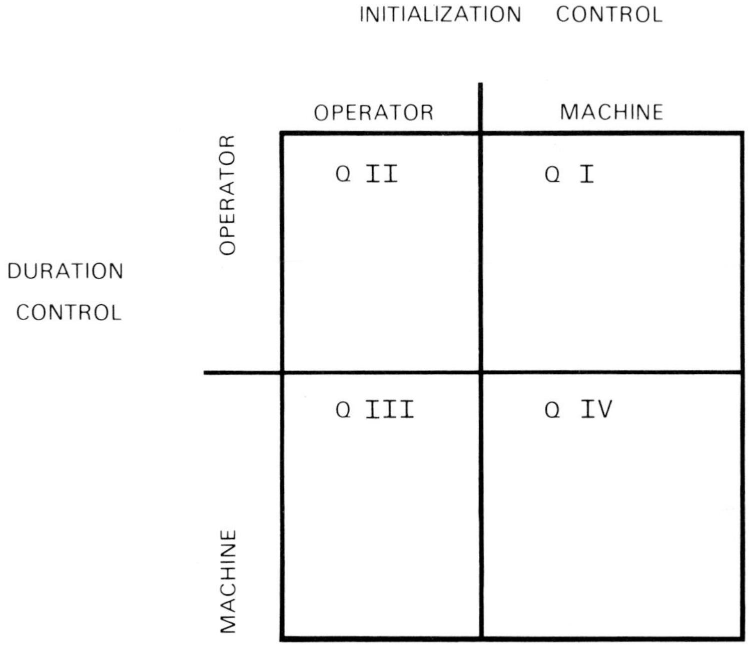

INITIALIZATION   CONTROL

Figure 2. Classification system for paced work.

(Dainoff 1979) suggests the need for the investigation of effects of this type of work.

Quadrant III (QIII) tasks are initiated by the operator, but the machine determines the work time. QIII tasks appear to be similar to those that Murrell (1963) described as Type 1. He states, 'This type of pacing is found when girls feed machines with parts which simply have to be picked up from a bin and placed in an appropriate position . . . ' The operator feeds parts into a waiting machine; the machine then processes it while the operator is prevented from loading another part.

The majority of work has been done in the quadrants labelled II and IV. Quadrant II (QII) describes tasks that are often referred to as unpaced (e.g., Conrad 1955, Dudley 1962). The operator both initiates a QII task and controls the length of time to complete it. QIV tasks are often referred to in the literature as machine paced (e.g., Conrad 1955). The machine (or external environment) both initiates a QIV task and controls the length of time to complete it.

The usefulness of the above system resides not only in its comprehensiveness, but in the fact that the basis of its classification rests on a concept which has been demonstrated to be a potent predictor of occupational stress. Cross-quadrant comparisons—in both field and laboratory investigations—would seem particularly fruitful as a means of elucidating the precise role of operator control as a determinant of stress responses. Furthermore, addition of a third dimension to figure 2 would allow independent specification of job demand level (Karasek 1979) within each quadrant.

References

BULEY, G. M., SLACK, N. D., and WILD, R., 1973, Production flow line system design—a review. *AIIE Transactions*, March, 18.

CONRAD, R., 1954, The rate of paced man-machine system. *Journal of the Institution for Production Engineering*, **33** (10), 562-570.

CONRAD, R., 1955, Comparison of paced and unpaced performance at a packing task. *Occupational Psychology*, **29**, 19.

CONWAY, E., STREIMER, I., SANDERS, M. S., and RIEMER, S. E., 1977, *Search for Health and Safety Data on Machine-Paced Assembly Lines*, Contract No. HMS-210-76-0148, US Department of Health, Education and Welfare, 19.

DAINOFF, M., 1979, *Occupational stress factors in secretarial/clerical workers* (Washington: National Institute for Occupational Safety and Health).

DUDLEY, N. A., 1962, Work time distributions. *Internation Journal of Production Research*, **2**, 20.

FRANKS, I. T., 1974, Some characteristics of conveyor-based operator performance. *Work Study and Management Services*, **18** (7), 20.

HAPP, A., 1981, *Effects on an unpaced/paced video-coding task on performance and heart rate indicators*. Unpublished M. A. Thesis, Department of Psychology, Miami University, Oxford, Ohio, USA.

HURRELL, J. J., Jr, 1981, *Psychological, physiological and performance consequences of paced work. An integrative review* (Cincinnati, Ohio: National Institute for Occupational Safety and Health).

KARASEK, R. A., Jr, 1979, Job demands, job decision latitude and mental strain: implications for job redesign. *Administrative Science Quarterly*, **24**, 285-370.

MURRELL, K. F. H., 1963, Laboratory studies of repetitive work I: Paced work and its relationship to unpaced work. *International Journal of Production Research*, **2** (3), 21.

ROHMERT, W., and LUCZAK, H., 1973, Zur ergonomischen Beurteilung informatorischer Arbeit. *International Zeitschrift fur angewandte Physiologie einschliesslich Arbeitsphysiologie*, **31**, 209-229.

SURY, R. J., 1967, The simulation of a paced single-stage work task. *International Journal of Production Research*, **4** (2), 22.

The human being and paced work on the assembly line

By V. Bulat

The Faculty of Mechanical Engineering, Department of Industrial and Management,
27. marta 80, 11000 Beograd, Yugoslavia

The results of an investigation of assembly-line work show that the structure of the work is similar to that in other countries. Nearly 90% of the assembly-line workers regard their work as being monotonous. The alienation could be alleviated as follows: (1) the participation in self-management decision-making and discussion on current production problems should be part of the working day; (2) there should be four or five extra short stops for rest; and (3) the environment around the assembly line should be optimized.

1. Introduction

The human being at work must be thoroughly studied. Underdeveloped multidisciplinary collaboration slows down the progress in this area.

The industrial transformation of the developing countries is carried out through technological transfer from the developed countries. The introduction of the assembly line occupies an important position in this transfer. Yugoslavia introduced the assembly-line system, in addition to some other results of technological progress, during its postwar industrialization process. Yugoslavia's specific circumstances mean that a transformation of the socio-economic system, based on self-management principles, has been going on parallel with the industrialization process. Essentially, self-management permeates through the lives and work of every member of the community. The assumption that self-management potentialities should compensate for the negative aspects of the assembly-line system in relation to man as an integral human being seems to be well grounded. This text concentrates on the presentation of the methodological approach and characteristic results obtained until now.

2. Investigation methodology

The fact that before the research of Bulat *et al.* (1976) there had been no systematic study of the assembly-line system in Yugoslavia determined the approach to the investigation to a great extent. The starting hypothesis opened some questions connected with the rate of assembly-line distribution in industry and with the general characteristics of installed assembly lines, for the purpose of evaluating the nature and relevant aspects of the existing problems. The next step, depending on the outcome of the first one, included a many-sided study of industrial assembly-line systems.

Consequently, there was a need for designing two mutually conditioned research procedures: (1) general or diagnostic procedure; (2) specific procedure, covering a complex scope.

An examination of the available sources of information on the assembly-line study (Walker and Guest 1952, Friedman 1959, White 1966, Aguren *et al.* 1976) showed that there had been no projects developed with similar aims. However, what was developed by the above authors regarding research methods and techniques proved to be of fundamental importance in the design of these investigations.

An emphasis was laid on questionnairing and data recording, while interviewing was only used to supplement and check information collected through questionnaires and recordings. The data recording and questionnairing was conducted by a number of research assistants.

A very interesting research undertaking was conducted by Runcie (1980), in which the author, as an anonymous workman on the assembly line for several months, was trying to avoid possible deformations usually arising from direct communication between the workman and the leading researcher. The procedure applied in this case tends to prevail over the distinct elements of subjectivity.

To conduct the questionnairing and data recording, two groups of requirements were designed: (1) general questionnaire; and (2) complex questionnaire.

The general questionnaire was sent out to 900 factories which were assumed to possess an assembly-line system. Replies of a predominantly satisfactory quality of information returned from 100 factories.

The complex questionnaire was also used to collect, through questions, observation, data recording and interviews, the data required for the project.

Eight factories and 250 workers on assembly lines were included in the pool. The plants belonged to the metalworking, food-processing, clothing and shoe-making industries. The object of investigation in each of the factories was their representative assembly line.

The documentation base, which was formed subsequently, contains qualitatively and quantitatively satisfactory data.

3. Investigation results and discussion

The processing of available documentation from the general questionnaire made it possible, among other things, to investigate the following: (1) assembly-line distribution through the percentage of men working on assembly

Table 1. Characteristic indices of the personnel structure.

Group of workers	Rate level	Sex	Percentage %	Percentage M and F workers together
Workers on assembly lines with age	Up to 20 years	M	24·52	13·96
		F	75·47	
	20 up to 30 years	M	29·36	55·83
		F	70·63	
	Over 30 years	M	35·56	29·01
		F	64·43	
Workers on assembly lines with practical experience	Up to 20 years	M	29·17	23·83
		F	70·90	
	2 up to 5 years	M	34·32	29·82
	Over 5 years	M	32·73	46·34
		F	67·27	
Workers on assembly lines with level of skill	Unskilled	M	27·33	0·63
		F	72·66	
	Semi-skilled	M	28·86	61·27
		F	71·13	
	Skilled	M	38·56	37·80
		F	61·43	
	High skilled	M	68·42	0·15
		F	31·57	
	High school	M	21·87	0·13
		F	78·12	

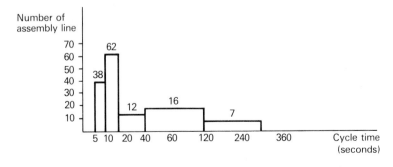

Figure 1.

lines in the industries polled; (2) characteristic indices of the personnel structure (table 1); and (3) pace duration (figure 1).

On the whole, $41 \cdot 8\%$ of the total number of workers engaged in productive labour in the factories work on assembly lines.

The structure of assembly-line workers is characterized by the following facts: (1) female labour is predominant; (2) younger workers up to 30 years of age are substantially represented ($69 \cdot 79\%$); and (3) by far the largest group in number are semi-skilled workers ($61 \cdot 27\%$).

From the above rates one can conclude that the structure of assembly-line workers reflects the usual picture of the same structure in other countries. This structure could have been expected since technology transfer is accompanied by transfer of organization.

Figure 1, showing the pace duration, is indicative of an important characteristic of the assembly lines included in the investigation. Four-fifths of the assembly lines have a pace duration up to 60 seconds. A limited pace duration contributes to elements of the worker's impoverished commitment, which leads to alienation.

The above and some other results of the first stage of this investigation indicate the existence of conditions for the assembly-line worker's alienation. The second stage should make it possible, on the basis explained here, to interpret and penetrate more deeply the alienation problem.

The naturally stable and rhythmical work causes monotony and other inconveniences among assembly-line workers. The examined assembly lines show a different state, characterized by the following.

(1) Working tasks have a stable content, and periodic changes in the type of product do not result in a relevant change in work content.

(2) The pace duration ranges from 15 s to 150 s.

(3) Rhythmical repetition of identical work contents was designed for all assembly lines, but due to some organizational weaknesses the work percentage that could be recorded ranges between $40 \cdot 73\%$ and $81 \cdot 7\%$, the greater part being grouped around 80% of work.

It follows from the above data that technological and organizational conditions are such that negative effects can be expected in the workers' reactions.

The ecological factors of the working environment, and the aesthetic shaping of the space around assembly lines are characterized by the following.

(1) Workshop temperatures are within the bounds of 19 °C and 32 °C, but in six cases temperatures of 24 °C or higher were registered, which is considerably above the temperature recommended for this type of work.

(2) Humidity of the air ranges from 29% to 60%, which is quite satisfactory considering the fact that humidity is generally lower in shops subjected to higher temperatures.

(3) Noise ranges between 70 dB and 94 dB, which touches the upper limit or exceeds the allowable value.

(4) Air pollution exceeds the allowable limits on five assembly lines.

(5) Except for one example, the aesthetic shaping of the workshop space is neglected.

Reduction and elimination of causes of the qualitative degradation of ecological conditions is considered to be a significant step towards mitigating the negative effects of the assembly line upon the worker.

The polled workers' replies to the question of fatigue, such as 'Do you feel tired after work?', are 'great fatigue' in 79·5% of cases.

According to labour regulations, all workers are entitled to a 30 minute break for a hot meal within the eight-hour working day. The investigation shows that in four cases the workers had no additional stops for rest on their lines, while in the remaining four cases the situation is as follows: (1) on three assembly lines the workers have two additional rests lasting ten minutes each; and (2) on one of the lines the workers also have two additional rests, the duration being 20 and 10 minutes respectively.

Considering that nearly 90% of the polled workers regard their work as being monotonous (Parežanin 1979), some experiments were conducted on an assembly line to determine an adequate arrangement and duration of rests. The results showed that a suitable four-rest scheme would be the 10-20-10-10 minute schedule. On another assembly line, Gavrić (1980) experimentally determined the schedule to be 5-5-10-5-5. It is worth noting that the daily output is maintained in both cases. Consequently, three or four short rests, the middle one being twice as long, offer an optimum structure of stops for rest on assembly lines.

In the context of studying the effects of self-management on assembly-line workers, the following opinions expressed by the workers may be characteristic.

To the question, 'If you have stated your opinion about the negative consequences of assembly-line work, do you expect that the consequences may be eliminated or substantially reduced under self-management conditions?' 77·5% workers gave a positive answer. However, the answer to the question, 'Do you think you can exert any influence on decision-making in your organization?' are classed as follows:

I have an influence, 13·5%.
I have a partial influence, 20·6%.
I have a minor influence, 19·5%.
I have no influence, 46·5%.

This structure of workers' replies can be explained by the fact that self-management decisions are reached at all workers' meeting held during or after factory hours, but assembly-line workers cannot readily take part since in the former case that would cause downtime problems, while the reason in the latter case is their fatigue. Coming to self-management decisions and other agreements connected with production is a component of assembly-line workers' activities and therefore should be an integral part of their work tasks and, consequently, of their working hours. This would make self-management a system that allows complete self-expression of the assembly-line worker, while excluding the remaining elements of alienation.

4. Conclusion

The investigation conducted on the basis of narrowing and deepening a study of paced work on the assembly line make it possible to adequately evaluate some specific problems. The partially completed comprehensive study enables one to come to certain conclusions.

Firstly, participation in self-management decision-making, enriched with agreements on current production problems, should be an integral part of the workers' engagement within their working hours.

Secondly, assembly-line workers should be allowed an additional four or five stops for rest, three or four short ones and a longer one in the middle of factory hours, between the short rests.

Thirdly, ecological factors and aesthetic space shaping around the assembly line should be optimized.

Further investigations are expected to result in new measures tending to put an end to man's alienation at assembly-line work. Moreover, the possibility of replacing the assembly line with a new organizational-technological concept must not be excluded.

References

AGUREN, S., HAUSSON, R., and KARLSSON, K. G., 1976, *The Volvo Kalmar Plant (The Impact of New Design on Work Organization)* (Stockholm: The Rationalization Council SAF-LO).

BULAT, V. P., 1976, *i saradnici, 1976, Istraživanje ljudskog faktora u radu na proizvodnim trakama u industriji* (Beograd: Institute techničkih nauka SANU, Centar za ergonomska istraživanja).

FRIEDMAN, G., 1959, *Kuda ide ljudski rad* (Beograd).

GAVRIĆ, L., 1980, *Istraživanje funkcionisanja proizvodnih traka sa aspekta mogućnosti prilagodavanja rada radniku* (Beograd: Mašinski fakultet, magistrarski rad).

PAREŽANIN, V. J., 1979, *Istraživanje funkcionisanja proizvodne trake* (Beograd: Mašinski fakultet, magistarski rad).

RUNCIE, J. F., 1980, By days I make the cars. *Harvard Business Review,* Vol. 58, Number 3, 106–115.

WALKER, C. R., and GUEST, R. H., 1952. *The Man on the Assembly Line* (Cambridge, Mass: Harvard University Press).

WHITE, W. F., 1966, *Čovek i rad* (Zagreb: Panorama).

Work repetition and pacing as a source of occupational stress

By B. Beith

North Carolina State University, Department of Psychology,
Ergonomics Program, Raleigh, North Carolina 27607, USA

Work repetition and pacing have become increasingly common practices in industry for enhancing production efficiency. Questions have long been raised concerning potentially harmful stress effects of such job characteristics on the worker. A review of literature reveals indications of such stress; however, these have not corresponded specifically to any widespread ramifications evidenced in industry. This may be explained by the influence of system and individual moderators, as well as shortcomings in the literature and methodologies.

1. Introduction

Since the turn of the century, assembly-line processes involving workpacing and repetitive tasks have become more widely used in industry. It is estimated that in the USA alone over three million workers perform repetitive or paced jobs. Many researchers (Conrad 1954, Dudley 1962, Belbin and Stammers 1972) contend that neither industry nor, especially, the worker benefit from the potential stress on the worker under such work conditions. A review of the literature was undertaken to determine whether indications of stress existed and, if so, whether the ramifications of such stress were apparent in industry.

2. Indications of stress

There are indications in the literature that work repetition and pacing are potentially stressful to workers as evidenced by psychological, physiological, and performance indices.

Psychological indications include lowered levels of arousal and increased levels of perceived stress. Research indicates that repetitious work is associated with lowered arousal (Cox 1980) and that paced work results in lower arousal than unpaced work (Franks and Sury 1966, Manenica 1977). Knight *et al.* (1978) attribute increased arousal of unpaced work to a higher cognitive load on the worker to 'regulate' his own pace. Lowered arousal can lead to increases in 'attention shifts' by the worker from the task. Johansson *et al.* (1976) attribute this increase to the worker's attempt to maintain or increase arousal levels.

Pacing is also associated with increased levels of perceived stress (Cox and Mackay 1979, Ferguson 1973, Johansson *et al.* 1976). In a comparison of repetitious and paced work, Mackay *et al.* (1979) reported a decrease in arousal for both work conditions, but an increase in perceived stress as the pacing increased. After interviewing 205 assembly-line workers, Broadbent and Gath (1979) reported increased anxiety associated with pacing but not repetition *per se.* Cox (1980) interpreted increased perceptions of stress as another attempt to maintain arousal levels and offset monotony.

The physiological indications of stress are reflected by differences in catecholamine levels (in blood and urine), cardiac activity, and respiratory activity of workers under paced and unpaced conditions.

The most consistent and reliable indications of stress have been increased catecholamine levels of workers in repetitive, paced jobs (Johansson *et al.* 1976, Frankenhaeuser and Gardell 1976). Such increases indicated mobilization of the body to offset stress.

Cardiac-related activity is a more prevalent but equivocal indicator of stress in the literature. While some studies report increased heart rates (Amaria 1974) and higher blood pressure (Hokanson *et al.* 1971) for paced work compared to unpaced work, other studies found no such difference (Salvendy and Humphreys 1979). Cardiac R-R intervals reportedly change more rapidly during paced work (Manenica 1977) but are shorter during unpaced effort due to increased cognitive load (Salvendy and Humphreys 1979).

Generally, respiratory measures do not seem to be sensitive to industrial-type pacing tasks (Manenica 1977), although Salvendy (1972) reported higher activity for paced-exercise tasks when compared to unpaced effort.

Energy consumption curves have shown the greatest work efficiency associated with unpaced work, whether determined by cardiovascular (Corlett and Mahadeva 1970) or respiratory (Streimer 1969) indices.

Performance indications of stress under paced conditions include slower response times (Murrell 1963), shorter 'actile' periods of optimum performance (Murrell and Forsaith 1963), increases in irregular, erratic, or non-work-related movements (Salvendy 1972, Basila *et al.* 1979), and increased errors (Sury 1964, Salvendy and Humphreys 1979, Salvendy 1972). These findings seem to provide fairly consistent indications of performance decrement as a function of stress arising from repetitious or paced-work conditions.

3. Ramifications of stress

The ramifications of stress resulting from work repetition and pacing represent the impact of such stress in industry. In contrast to the research providing indications of stress, the ramifications of that stress have received little or no attention.

There was little reliable evidence of long-term physical health effects of repetition or pacing. Although studies reported worker complaints of angina, headaches, slight nervous disorders, and low-back disorders ·(Cooper and Marshall 1976, Johansson *et al.* 1976, Ferguson 1973, Teiger and Laville 1972), it is not clear whether these problems stem exclusively from repetition and pacing or from other factors such as workplace design.

Long-term mental-health problems are even less evident in the literature. Although researchers have suggested etiological influences of repetition and pacing on mental illness (Cox 1980, Broadbent and Gath 1979), no research support is apparent in the literature.

Cooper and Marshall (1976) point out that while there is sufficient indirect data indicating the danger of pacing to mental and physical health, there is insufficient evidence to draw firm conclusions. It should be noted that this same conclusion applies to accident/injury research relative to work repetition and pacing.

Job satisfaction of workers has received considerable attention but conclusions are equivocal.

While most research reports increased dissatisfaction for workers in repetitious or paced jobs (Coetsier 1966, Sanders *et al.* 1979), other studies report no such increase (Teiger and Laville 1972, Khaleque 1979).

The ramifications of stress due to work repetition and pacing are not dramatically evident in the literature. The question remains as to why.

4. Moderators of stress

Individual and system moderators may provide one explanation of why indications of stress but not ramifications of stress are evidenced in the literature.

Individual moderators cited in the literature as influencing the effects of work repetition and pacing are sex, age, and personality.

Female workers may have less pronounced physiological, but more pronounced psychological responses than male workers to such stress (Cox and Mackay 1979). Although both were subjects in the studies, no comparisons were apparently made.

Early research concluded that older workers were less suited to paced work (Brown 1957); however, current studies report older workers to be more energy-efficient under paced conditions (Salvendy and Pilitsis 1971) and less prone to non-work-related movements (Basila *et al.* 1979) than younger workers.

Much of the variance among individual workers may reflect personality differences. The literature indicates that workers who are trusting, introverted, reserved, less intelligent, more tense, and possess a high tolerance for boredom are generally more suited to machine-paced work, whereas workers with opposite traits are more suited to self-paced work (Stagner 1975, Salvendy and Humphreys 1979, Sanders *et al.* 1979).

System moderators have received considerable attention in the literature. The most important moderator seems to be the tolerance of the system which allows worker flexibility to balance long and short cycles (Buffa 1961, Sury 1964). Pacing speed is important since too fast or too slow a speed can increase stress on a worker. High system tolerance can offset this somewhat (Sury 1964). Moderators such as work–rest cycles (Murrell and Forsaith 1963, Murrell 1963) and process-line length (Chase 1975) can influence a worker's ability to cope with repetition and pacing.

5. Shortfalls in the literature and methods

Aside from the influence of individual and system moderators in reducing the impact of stress in industry resulting from work repetition and pacing, shortcomings in the literature and methods may be obscuring those ramifications or providing inaccurate indications of stress. These shortcomings are as follows:

(1) Long-term effects on physical and mental health have not been adequately studied using longitudinal or cross-sectional designs. Thus, we have no point of reference with which to assess the ramifications of work repetition and pacing or to compare changes in the light of research.

(2) The influence of individual moderators has not yet been adequately studied. Further, few studies in the literature provide full ANOVA tables from which to determine the variance accounted for by repetition and pacing versus individual differences.

(3) Laboratory experiments do not always involve representative tasks. In most studies reviewed, less than ten subjects were used and for artificially short periods of time (e.g., less than two hours).

(4) There appears to be little correspondence between measures and task situations in the laboratory and the field. Validation of laboratory research through field efforts could improve generalizability and lead to intervention research where needed.

The use of repetition and pacing to enhance efficiency and productivity has become a postulate of progress. The resulting stress, however, may cost the worker and industry both. It is effective research into the indications, ramifications, and moderators of such stress that can help accomplish the goals of industry while upholding the right of the worker to health and safety.

References

AMARIA, P. J., 1974, Effect of paced and unpaced work situations on the heart rate of female operators. In *Development of Production Systems*, edited by C. H.

Gudnason and E. N. Corlett, Proceedings of the Second International Conference at Copenhagen, Denmark, 27–31 August (London: Taylor & Francis).

BASILA, B., SUOMINEN, S., SALVENDY, G., and McCABE, G. P., 1979 Non-work related movements in machine paced and in self-paced work. *Proceedings of the 23rd Annual Meeting of the Human Factors Society*, Boston, Massachussets, 29 Oct. – 1 Nov.

BELBIN, R. M., and STAMMERS, D., 1972, Pacing stress, human adaptation and training in car production. *Applied Ergonomics*, **3** (3), 142–146.

BROADBENT, D. E., and GATH, D., 1979, Chronic effects of repetitive and non-repetitive work. In *Response to Stress Occupational Aspects*, edited by C. Mackay and T. Cox (Guildford: IPC Science and Technology Press).

BROWN, R. A., 1957, Age and paced work. *Occupational Psychology*, **31**, 11–20.

BUFFA, E. S., 1961, Pacing effects in production lines. *Journal of Industrial Engineering*, Nov. – Dec., 383–386.

CHASE, R. B., 1975, Strategic considerations in assembly-line selection. *California Management Review*, **28**, (1), 17–23.

COETSIER, P., 1966, An approach to the study of the attitudes of workers on conveyor belt assembly lines. *International Journal of Production Research*, **5** (2), 113–135.

CONRAD, R., 1954, The rate of paced man–machine systems. *Journal of the Institution of Production Engineers*, **33**, October, 52–58.

COOPER, G. L., and MARSHALL, J., 1976, Occupational sources of stress: a review of the literature relating to coronary heart disease and mental ill health. *Journal of Occupational Psychology*, **49**, 11–28.

CORLETT, E. N., and MAHADEVA, K., 1970, A relationship between freely chosen working pace and energy consumption curves. *Ergonomics*, **13** (4), 517–524.

COX, T., 1980, Repetitive work. In *Current Concerns in Occupational Stress*, edited by C. L. Cooper and R. Payne (New York: John Wiley).

COX, T., and MACKAY, C. J., 1979, The impact of repetitive work. In *Satisfactions in Job Design*, edited by R. Sell and P. Shipley (London: Taylor & Francis).

DUDLEY, N. A., 1962, The effects of pacing on worker performance. *International Journal of Production Research*, **1** (60), 137–149.

FERGUSON, D. A., 1973, Comparative study of occupational stress. *Ergonomics*, **16**, 649–664.

FRANKENHAEUSER, M., and GARDELL, B., 1976, Underload and overload in working life: Outline of a multidisciplinary approach. *Journal of Human Stress*, **2** (3), 35–46.

FRANKS, I. T., and SURY, R. J., 1966, The performance of operators in conveyor-paced work. *International Journal of Production Research*, **5** (2), 97–112.

HOKANSON, J. E., DEGOOD, D. E., FORREST, M. S., and BRITTAIN, T. M., 1971, Availability of avoidance behaviours in modulating vascular-stress responses. *Journal of Personality and Social Psychology*, **19**, 60–68.

JOHANSSON, G., ARONSSON, G., and LINDSTRÖM, B. O., 1976, *Social psychology and neuroendocrine reactions in highly mechanized work*. Report from Department of Psychology, University of Stockholm, No. 488.

KHALEQUE, A., 1979, Performance and job satisfaction in short-cycled repetitive work. In *Satisfaction in Job Design*, edited by R. Sell and P. Shipley (London: Taylor & Francis).

KNIGHT, J. L., SALVENDY, G., ENDICOTT, G., BASILA, B., and SHARIT, J., 1978, Locating and reducing cognitive load in paced and unpaced tasks. *Proceedings of the 22nd Annual Meeting of the Human Factors Society*, Detroit, MI, October 16–19.

MACKAY, C. J., COX, T., WATTS, C., THIRLAWAY, M., and LAZZERINI, A. J., 1979,

Psychophysiological correlates of repetitive work. In *Response to Stress Occupational Aspects*, edited by C. Mackay and T. Cox (Guildford: IPC Science and Technology Press).

MANENICA, I., 1977, Comparison of some physiological indices during paced and unpaced work. *International Journal of Production Research*, **15** (3), 161–175.

MURRELL, K. F. H., 1963, Laboratory studies of repetitive work I: Paced work and its relation to unpaced work. *International Journal of Production Research*, **2** (3), 169.

MURRELL, K. F. H., and FORSAITH, B., 1963, Laboratory studies of repetitive work II: Progress report on results of two subjects. *International Journal of Production Research*, **2**, 247.

SALVENDY G., 1972, Paced and unpaced performance. *Acta Physiologica*, **42**, 267–274.

SALVENDY, G., and HUMPHREYS, A. P., 1979, Effects of personality, perceptual difficulty, and pacing of a task on productivity, job satisfaction, and physiologic stress. *Perceptual and Motor Skills*, **49**, 219–222.

SALVENDY, G., and PILITSIS, J., 1971, Psychophysiological aspects of paced and unpaced performance as influenced by age. *Ergonomics*, **14**, 703–711.

SANDERS, S., SALVENDY, G., KNIGHT, J. L., and McCABE, G. P., 1979, Attitudinal, personality, and age characteristics for machine-paced and self-paced operations. *Proceedings of the 23rd Annual Meeting of the Human Factors Society*, Boston, Massachussetts, Oct. 29–Nov. 1.

STAGNER, R., 1975, Boredom on the assembly line: age and personality variables. *Industrial Gerontology*, **2**, 23–44.

STREIMER, I., 1969, Considerations of energy investment as determinants of behaviour measurements. *Proceedings of the International Ergonomics Association Conference on Men and Machines*, Amsterdam, Sept.

SURY, R. J., 1964, An industrial study of paced and unpaced operator performance in a single stage work task. *International Journal of Production Research*, **3** (2), 91.

TEIGER, C., and LAVILLE, A., 1972, The nature and variation of mental activity during repetitive tasks: an attempt to evaluate workload. *Le Travail Humain*, **35** (1), 99–116.

Perceptual narrowing and forced-paced tasks

By T. J. TRIGGS

Department of Psychology, Monash University,
Clayton, Victoria 3168, Australia

Human operators under task-induced stress conditions are frequently regarded as undergoing a 'narrowing' of their visual perceptual field. Typically, the effect has been demonstrated by the decrements in responding to peripheral visual signals as the task demands of a central or primary task are increased.

This paper discusses the criteria that must be satisfied before the phenomenon of perceptual narrowing can be deemed to have occurred. These are based on the statistical requirement to show a significant interaction between the factors of visual angle and task demand for performance on the peripheral task, and the need to avoid floor or ceiling effects. Data from several experiments are presented using forced-paced central tasks with different input rates that argue against the generality of the concept of perceptual narrowing.

1. Introduction

One frequently claimed manifestation of stress is that of a narrowing of a human operator's visual field. This phenomenon has been referred to as 'perceptual narrowing' (Weltman and Egstrom 1966), 'tunnel vision' (Mackworth 1965), or 'funnelling' (Bursill 1958). Such phenomena have been associated with different forms of environmental, emotion and task-induced stress.

The generality of the concept of perceptual narrowing will be challenged here. The criteria necessary to support strong claims of narrowing will be considered. Previous studies dealing with the effect of the task-induced stress resulting from performance of a central task, usually visual, on responding to simple signals spread across the visual periphery will be discussed briefly. These studies overall do not provide unqualified support for the concept of narrowing under task-induced stress. A procedure will be suggested for studying peripheral performance so as to allow perceptual narrowing to occur unaffected by other factors that could influence performance.

Three sets of experimental data gathered in a laboratory setting will be presented which examined peripheral performance as a function of externally-paced central tasks for a range of input rates. These studies used either detections or reaction times for recording peripheral performance. Their aim was the evaluation of the perceptual narrowing phenomenon and the assessment of the effect of central-task pacing on peripheral performance.

2. Theoretical considerations

Some researchers have made a reduction in the range of the perceptual stimuli used by subjects under stress an integral part of their theoretical considerations (e.g., Easterbrook 1959, Teichner 1968). Such a reduction is said to have occurred when the use of peripheral cues is reduced in the presence of stress. For task-induced stress, degradation of this type is typically attributed to the involvement of central attentional processes, because it is not found when the processing load on the subject is removed while other task characteristics are maintained.

Some authors have discussed such a reduction in terms of a shrinkage of the limits of the effective visual field (Mackworth 1965). Such a strong model of a narrowing effect implies an actual change in the size of the functional visual field, while peripheral stimuli within the limits of the field are perceived unchanged. Other less extreme terms, such as funnelling or narrowing, suggest a phenomenon where higher central-task load causes a larger decrement in peripheral performance as the angle of eccentricity is increased. This latter description is the one being considered here.

3. Criteria of assessment

When the load-induced stress of a central task is added or increased, some investigators have found a systematic reduction in the level of performance in responding to the peripheral signals that was *greater* the more peripheral the light position (Bursill 1958, Youngling 1965). However, our re-analysis of data from other relevant studies indicated that such a steady increasing trend in decrement at greater angles of eccentricity did not always occur. Experiments showing an increase in either absolute or percentage terms include Moskowitz and Sharma (1974), and Holmes *et al.* (1977). Those that do not show a consistent increasing trend include Webster and Haslerud (1964), Leibowitz and Appelle (1969), Bartz (1976), and Weltman *et al.* (1971).

Some authors have claimed that perceptual narrowing has been demonstrated when peripheral performance overall is degraded. This does not of itself provide support for the concept. What is needed is a differential decrement, greater in the far visual periphery than the near, which is not a function of instructions to the subject, either explicit or implicit. For example, it is possible to interpret the results of Bahrick *et al.* (1952) in terms of an attentional trade-off between the central and peripheral tasks caused by the instructions rather than in terms of narrowing. Furthermore, one must guard against the possibility of other performance factors, such as floor/ceiling effects or trade-offs within the peripheral task itself, influencing performance because of the perceived demands of the situation. Almost invariably in studies in this area, because equal intensity signals are used across the periphery, performance in the control condition has been considerably worse in the far periphery than the near. Under the stress conditions, this introduces the strong likelihood of differential degradation occurring because of spurious factors rather than from the operation of a mechanism that narrows the functional visual field or selectively degrades the far periphery.

One approach that has received little attention is that of equating performance across the periphery, at least approximately, in the control condition. If any spurious performance factors were to play a role, then conditions should be equivalent at all angles of eccentricity. If perceptual narrowing is a robust phenomenon, then far peripheral performance decrements should exceed any that are found in the near periphery when a centrally placed task is added or its performance demands increased.

Relevant data from experiments designed in collaboration with P. N. J. Lee and H. M. Khalid as part of their Master's programme research will be briefly reported to illustrate the application of the performance-equating approach. In the first of these studies, the detectability of signals across the visual field was approximately equated using pilot subjects. In the remaining two experiments, the intensity of signals was adjusted so that subjects responded with nearly equal reaction times to each peripheral position. All data were collected using a forced-paced central task.

4. Experiment 1

4.1. *Method*

In this laboratory study, subjects reported the presence of peripheral signals with and without a four-choice central task. Twelve subjects performed in the experiments, responding to low-intensity peripheral lights mounted at $20°$, $40°$, $60°$ and $80°$ to the left and right of the central position. The intensity of the lights was adjusted to yield approximately equal detection rates at each light location. Only one peripheral light was presented at a time at the rate of 12/min with approximately random interstimulus intervals. In the central task, stimuli were presented at the rate of 52/min. Each subject performed on the central numbers alone, peripheral lights alone, and the two together for a total of 15 two-minute blocks.

4.2. *Results*

All light positions had fewer detections when performance on the central task was required, as shown by the plot of detection decrement for each peripheral angle shown in figure 1. However, the interaction between presence/ absence of demand and peripheral angle was not statistically significant (F(1,11) = 0·98, *p*⟨0·10). Thus, there is no evidence of a reliable *differ-ence in decrement* across the light positions. However, in descriptive terms it tended to be larger at higher angles of eccentricity (with the exception of 80° left). Thus, while there was no statistical support for narrowing, the shape of the decrement curve was neither substantially flat nor concave-down, which would have provided strong evidence against narrowing.

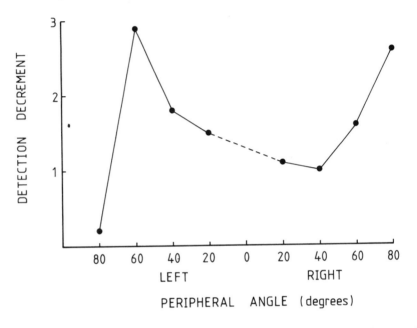

Figure 1. Decrement in the number of peripheral lights detected, averaged over sub-jects, as a result of adding the central forced-paced task for each peripheral light (Experiment 1).

Performance on the central task was significantly degraded when combined performance was required, but the decrement was quite small (about 3%).

5. Experiment 2

5.1. *Introduction*

While the statistical analysis in the previous experiment provided no support for narrowing under central-task load, the results fell short of being conclusive because of the shape of the curve. It was decided to examine the issue further in the following experiments by using reaction time (RT) as the prime depend-ent variable. This had several possible advantages. The type of measure on the central and peripheral tasks would be the same. It was also expected that the equal performance criterion across the periphery could be more easily approximated using this measure, and that RTs would be more sensitive to the experimental manipulations.

This experiment was designed specifically to evaluate the effect of the central-task input rate on peripheral performance. It could be argued that as the input rate to the central task was increased, the human information-processing system would have less capacity or time available to attend to the peripheral signals, so that increasing central-task rate would degrade peripheral performance. Alternatively, it is conceivable that subjects could allocate capacity in some proportion to both tasks and maintain this relative emphasis while the input rate was altered. Such capacity allocation has received support in other types of multiple-task performance (Triggs 1968). In this case, peripheral performance should be largely independent of the pacing of the central task.

5.2. *Method*

The central task was similar to that of Experiment 1, but used input rates of 48 (slow), 60 (moderate), and 72 (fast) per minute. The peripheral lights were the same as those of Experiment 1, except that an additional light was added centrally under the digit display. Pilot subjects were used to establish an

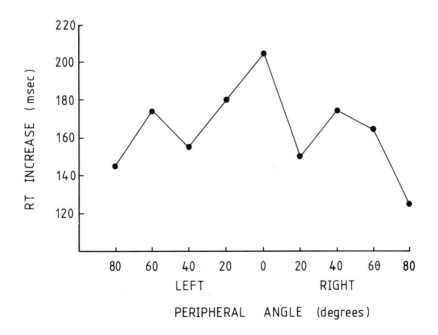

Figure 2. Decrement in peripheral RT, averaged over subjects, as a result of adding the central forced-paced task for each peripheral light (Experiment 2).

approximately equal RT across all nine lights, at a value just under 600 ms. In experimental trials, peripheral stimuli were presented at varying intervals at an average of 6/min. A total of 48 subjects were run in this experiment in two groups A and B, and each participated in the peripheral task only, central task only, and combined task conditions with three 3 min trials in each condition.

5.3. *Results*

The effect of adding the central task was to degrade the RT performance at each of the light positions as shown in figure 2. For the peripheral RTs, the interaction between presence/absence of the central task and angle of eccentricity was not statistically significant for either group A ($F(1,23) = 2 \cdot 78$, $p \langle 0 \cdot 10$) or group B ($F(1,23) = 3 \cdot 34$, $p \langle 0 \cdot 05$). This absence of significant interactions, and the shape of the curves with the smallest decrements tending to occur in the far periphery, argue strongly against the existence of perceptual narrowing in this type of task-induced stress situation. Changing stimulus input rate had no significant effect overall on peripheral performance ($F(1,46) = 3 \cdot 38$, $p \langle 0 \cdot 05$). Central task RTs were very similar for each of the three input rates.

6. Experiment 3

6.1. *Introduction*

This experiment represented the introduction of a different forced-paced central task, again using single digit stimuli but requiring contingent information processing. This task was introduced because it was considered more likely that central-task performance in this case would be influenced by input rate than the information conservation task used in the previous experiments. It could be argued that the failure to show an effect of rate in Experiment 2 resulted from the central task overall not being demanding enough. Two stimulus input rates were used in this experiment, namely 48 and 72/min, that corresponded to the slow and fast conditions in the previous experiment.

Another factor was introduced in this experiment. The effects of instructional trade-off on peripheral performance were evaluated. In previous experiments, the required attentional trade-off between the central and peripheral tasks has not been specifically investigated. It is conceivable that narrowing might only occur when the peripheral task is given a secondary role rather than an approximately equal or major emphasis. Apparent narrowing effects may have been shown in previous experiments because of the particular priority between tasks chosen by subjects, either as a direct result of the instructions given, or through assumptions made by subjects based on the perceived demands of the tasks.

6.2. *Method*

Two groups of subjects were run, with 12 subjects in each group. In both groups, two instructional conditions were investigated. In the first group, one of these gave priority to the central task, and the other required equal attentional emphasis to both tasks. For the other group, the instructional conditions were priority to the peripheral task and equal emphasis. The central task was of a contingent-processing type, and again RT performance was roughly equated across the periphery. Other aspects of the experiment were similar to the previous study.

6.3. *Results*

In the equal-emphasis condition, the addition of the central task again degraded RTs to the peripheral stimuli as shown in figure 3. The interaction between presence/absence of the central task and peripheral angle did not approach significance ($F(1,23) = 0 \cdot 90$, $p \langle 0 \cdot 10$). Figure 3 shows that the curve of the RT decrement was substantially flat, and these findings again argue against perceptual narrowing.

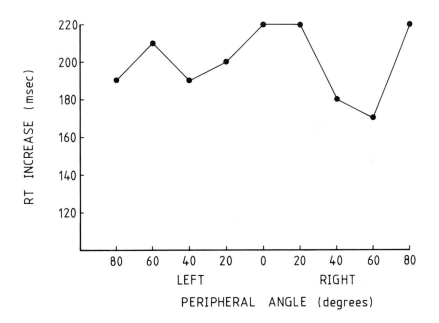

Figure 3. Decrement in peripheral RT, averaged over subjects, as a result of adding the central contingent forced-paced task for each peripheral light (Experiment 3).

Central-task rate did not influence peripheral RTs significantly in the equal-emphasis condition ($F(1,23) = 0 \cdot 36$, $p \langle 0 \cdot 10$). However, in contrast with the previous study, rate did influence performance on the central task.

Differential instructions produced the expected differences overall for peripheral performance, with the faster RTs being obtained when the peripheral task was given priority $(F(1,22) = 8 \cdot 00, \ p\langle 0 \cdot 01)$. The interaction between attentional emphasis and peripheral angle was not significant $(F(1,22) = 0 \cdot 56, \ p\langle 0 \cdot 10)$, while the RT curves were essentially flat for both attentional conditions. This argues against the proposition that perceptual narrowing will be shown if priority is given to the central task.

7. Discussion and conclusions

The use of the peripheral-performance equating approach in three experiments produced data that argue strongly against the concept of perceptual narrowing in task-induced stress situations. While such narrowing may be found under some conditions of physiological stress, the evidence here argues against the generality of the phenomenon.

While industrial-task designers can expect a general decrement in peripheral activities when the overall information-processing load is increased, marked degradation in the far periphery can be counteracted by an increase in the detectability or discriminability of the signals. While an increase in input rate may affect the central task, no evidence was found that such an increase in the amount of time required for the central task caused a decrement in the additional peripheral RT task. This was so whether attentional emphasis was given to the central task or not. The input rates in these experiments varied from 48 to 72 signals/min, which covers the range of interest in many industrial tasks. Thus, based on these results, it can be argued that under task-induced stress a capacity allocation strategy, which is not affected by changes of rate within reasonable limits, will be found in many applied situations. Such an allocation of capacity should depend on the perceived importance of task components and the overall task organization.

References

BAHRICK, H. P., FITTS, P. M., and RANKIN, R. E., 1952, Effect of incentives upon reactions to peripheral stimuli. *Journal of Experimental Psychology*, **44**, 400–406.

BARTZ, A. E., 1976, Peripheral detection and central task complexity. *Human Factors*, **18**, 63–70.

BURSILL, A. E., 1958, The restriction of peripheral visual during exposure to hot and humid conditions. *Quarterly Journal of Experimental Psychology*, **10**, 113–119.

EASTERBROOK, J. A., 1959, The effect of emotion on cue utilization and the organization of behaviour. *Psychological Review*, **66**, 183–201.

HOLMES, D. L., COHEN, K. M., HAITH, M. M., and MORRISON, F. J., 1977, Peripheral visual processing. *Perception and Psychophysics*, **22**, 571–577.

KHALID, H. M., 1981, *Peripheral visual performance and central task complexity*. Unpublished Master's thesis, Monash University.

LEE, P. N. J., 1976, *Driving task demand and peripheral visual detections.* Unpublished Master's thesis, Monash University.

LEIBOWITZ, H. W., and APPELLE, S., 1969, The effect of a central task on luminance thresholds for peripherally presented stimuli. *Human Factors,* 11, 387–392.

MACKWORTH, N. H., 1965, Visual noise causes tunnel vision. *Psychonomic Science,* 3, 67–68.

MOSKOWITZ, H., and SHARMA, S., 1974, Effects of alcohol on peripheral vision as a function of attention. *Human Factors,* 16, 174–180.

TEICHNER, W. H., 1968, Interaction of behavioural and physiological stress-reactions. *Psychological Review,* 75, 271–291.

TRIGGS, T. J., 1968, Capacity sharing and speeded reaction to successive signals. *Human Performance Center Technical Report 9,* Ann Arbor, Michigan, University of Michigan.

WEBSTER, R. G., and HASLERUD, G. M., 1964, Influence on extreme peripheral vision of attention to a visual or auditory task. *Journal of Experimental Psychology,* 88, 269–272.

WELTMAN, G., and EGSTROM, G. H., 1966, Perceptual narrowing in novice drivers. *Human Factors,* 8, 499–506.

WELTMAN, G., SMITH, J. E., and EGSTROM, G. H., 1971, Perceptual narrowing during simulated pressure-chamber exposure. *Human Factors,* 13, 99–107.

YOUNGLING, E. A., 1965, *The effects of thermal environments and sleep deprivation upon concurrent central and peripheral tasks.* Unpublished Doctoral dissertation, University of Massachusetts.

Stress in short-cycle repetitive work: general theory and an empirical test

By J. F. O'HANLON

Traffic Research Centre, University of Groningen, 9750 AA Haren, The Netherlands

Epidemiological research has suggested that repetitive work can be stressful, but owing to deficient methodology, and the multicausal/ variable-effect nature of that stress, the results were inconclusive. Recent use of catecholamine excretion rates as a non-specific index of occupational stress has provided a clearer indication of the relatively high level of stress present in workers employed in the most repetitive tasks. A theory was advanced to explain the occurrence of this stress as the result of (1) habituation, (2) diminished arousal leading to transient performance failures, (3) re-occurring compensatory effort, (4) anticipation of failure with associated feelings of anxiety and hostility. One prediction was made from this theory; i.e., in short-cycle, machine-paced work, stress effects would increase with the difficulty of the task's perceptual requirement. This was confirmed in a laboratory simulation of a repetitive packaging task.

1. Introduction

The purpose of this presentation is to review evidence indicating that short-cycle, repetitive work can be stressful; then to summarize a psychophysiological theory to explain why and under what circumstances stress occurs; and finally to take an example from a recently complete experiment showing how one prediction from that theory was confirmed.

2. Inconclusive epidemiological results

Results obtained from epidemiological and attitudinal surveys have led many to the conclusion that short-cycle, repetitive work can become the source of great dissatisfaction and even mental or physical disease (Caplan *et al.* 1975, Ferguson 1973, Gardell 1971, Gilbertova and Glivicky 1967, Kornhauser 1965, Laville and Teiger 1975, Martin *et al.* 1980, Nerell 1975, Samoilova 1971). Contradictory findings have been obtained from other surveys (Roman and Trice 1972, Siassi *et al.* 1974) and the resolution of the ensuing argument has been impeded by the recognized inadequacy of the retrospective, cross-sectional survey approach employed by every investigator (see Kasl 1978).

Nonetheless, those who believe that repetitive work is stressful have at least convinced governing bodies in several nations to the extent that highly repetitive machine-paced work is severely restricted (e.g. in Norway) or will become so given the enactment of proposed legislation (e.g. in Sweden).

This trend is certainly in advance of the data available from epidemiological studies. Very few attempts have been made to isolate specific physical or organizational factors responsible for adverse reactions to repetitive work, and few of these were successful. (Exceptions: Saito 1973, Saito *et al.* 1972, Saito and Endo 1977.) Moreover, no one has been able to identify a simple chain of causal events leading from exposure to supposed occupational stressors (physical or psychological) through pathogenic reactions as mediated by individual, social and cultural variables, to final system breakdowns responsible for disease. It now seems likely that stress in repetitive work arises from the action of many simultaneous factors; that individual tolerance to their efforts varies widely; and that afflicted individuals react to their condition in a variety of adaptive and maladaptive ways (O'Hanlon 1981). Given that it is difficult, if not impossible, to identify specific causes and effects of stress in highly repetitive work, there is still hope of establishing at least a prima facie case for the pathogenicity of such work from measurement of a non-specific reaction of workers and comparison of that reaction among different occupational groups.

3. Inferring stress in repetitive work from catecholamine excretion

Most authorities now believe that the urinary excretion rates of catecholamine hormones, adrenaline and noradrenaline, adequately represent the sustained level of sympathoadrenomedullary activity in workers engaged in occupational tasks. Sympathoadrenomedullary activation is a classic non-specific reaction to stressors of all sorts; and theoretically, chronically elevated catecholamine production is directly responsible for a variety of psychosomatic diseases (Kagan and Levi 1974). Therefore, catecholamine excretion rates during work have become standard indices of occupational stress, allowing both a rough estimate of the total stress experienced by different workers in a particular occupation and a reasonable comparison of average stress levels in groups of workers from different occupations. A substantial data base now exists to facilitate such comparisons. Figure 1 summarizes some of these data from two sources (Åstrand and Rodahl 1977, Jenner *et al.* 1980).

Although this description is far from perfect, owing to occasionally small sample sizes and differences in analytical procedures, the results are certainly suggestive. Judging from catecholamine excretion rates, machine-paced assembly-line workers suffered the greatest stress.

Catecholamine excretion rates measured by Johansson *et al.* (1978) in sawmill workers provide an illuminating extension of this story. Workers

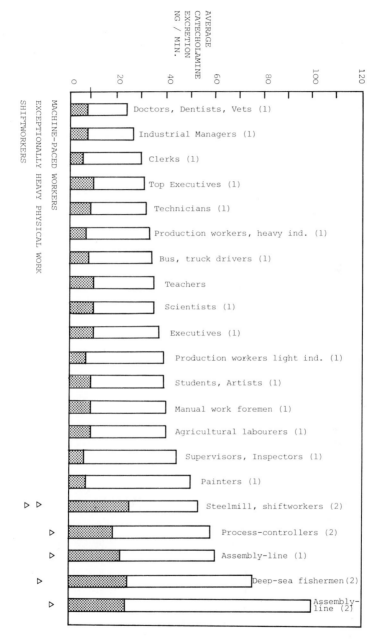

Figure 1. Comparison of catecholamine excretion rates during various types of work (Åstrand and Rodahl 1977, Jenner *et al.* 1980).

whose tasks were repetitive and machine-paced excreted adrenaline and noradrenaline at twice their resting rates, whereas those who performed several tasks at their own pace showed no catecholamine elevation. Adrenaline production was directly related to working pace, and inversely, to subjective wellbeing. Noradrenaline production was directly related to the degree of physical constraint imposed by the task, and also to feelings of irritation with the work.

From these findings we may more confidently conclude that highly repetitive work is indeed stressful. The question now becomes, why?

4. The genesis of stress in repetitive work: a theory

I recently attempted to relate the diverse subjective performance and biomedical reactions of workers who perform highly repetitive tasks within the context of a unifying psychophysiological theory (O'Hanlon 1981). I neglected the obviously important social, economic and cultural factors that mediate states of stress experienced by workers and focused instead upon what seemed the more fundamental, but not necessarily more important, reactions of the isolated individual to his task. Space here does not permit more than a superficial recapitulation of some of the more important conclusions and implications.

The more or less invariant requirement for repetitiously perceiving similar objects and responding to them in a stereotyped manner was seen as the necessary beginning of the stress state. Humans so treated do not easily maintain a level of higher mental activity conducive to efficient performance. Through the psychophysiological process of *habituation*, their level of cerebral cortical arousal is reduced below that which is optimal for efficient task performance. Given the fluctuating nature of arousal and task-directed attention, there would be no sudden or even monotonic change in the worker's level of performance efficiency. At the most, one can expect only a gradual increase in the variability of the worker's output or the quality of his performance. Eventually, however, the trend of diminishing arousal, attention, and performance efficiency would lead to a complete if transient performance failure. This would occur as arousal dipped through the minimum level required for the passage of enough information through the brain to execute the required motor activity in response to the perceptual signal. The classical term, 'blocking', was given by Bills (1931) to describe this phenomenon in repetitive laboratory tasks. Evidence that this also occurs in real tasks is available from a number of sources (e.g. Branton 1970, Haider 1962, Kogi 1972, Nilsson 1975, Stave 1977).

The individual's recognition of his brief performance failure occurs after the fact, though usually while there is still an opportunity to correct the error before the occurrence of a serious consequence. Nontheless, he now faces a difficult choice. Because the situation that brought on his failure has not changed, his failure to overcome its effects will permit repeated occurrences of similar blocking episodes and performance failures. In some real tasks and most tasks performed in laboratory experiments, performance failures will either go unnoticed by others, or if detected, will have no important consequences. But in other real tasks, repeated blocking episodes, of perhaps progressively longer duration, could be expected to result in profoundly important consequences—entire system failures apparent to everyone concerned, penalties, censure and even accidents involving injury (e.g. Branton 1970). He must therefore do something to counteract the process of habituation that is reducing his arousal and performance efficiency.

Workers in repetitive tasks apparently have recourse to a number of alternative behaviours which collectively may be considered as mechanisms for alleviating habituation or restoring arousal. Workers can sometimes pause, engage colleagues in conversation, busy themselves in another task such as machine checking or maintenance, etc. These subsidiary activities have all been observed to increase over time during repetitive work and do serve the function of maintaining measured arousal at a relatively high level in comparison to situations where rapid machine pacing prohibits their occurrence (Kishida 1973). In the latter situation, workers are seemingly left with the single recourse of maintaining their arousal by internal compensation through the self-initiated process of *effort*.

The scientific concept of effort originated largely from Kahneman (1973). To him, it was the capacity available to perform some task. The miscellaneous determinants of arousal in the task were said to determine some baseline level of effort. Beyond that, the voluntary mobilization of effort was possible to a limit that was in part intrinsic to the individual and in part set by external arousal determinants (i.e., more voluntary effort is possible in more arousing tasks). Kahneman's concept was modified by Pribram and McGuinness (1975), and again by Sanders (1979), to place effort in a compensatory feedback relationship with arousal. When arousal fails as the result of habituation, effort is the restorative process. There is no contradiction in the respective concepts. The capacity for voluntary effort may be reduced by a task-induced diminution in arousal (a positive feedback relationship) except when effort is elicited by the perception of blocking's consequences to restore arousal (a negative feedback relationship). This would explain why effort cannot indefinitely maintain arousal in repetitive work but how it might for a time restore arousal following a blocking episode.

Hockey (1979) recently observed that low arousal becomes stressful if an individual is required internally to counteract this state. Presumably he was referring to something like the process of effort as defined here. His opinion was supported by the recorded co-variance of subjective effort and stress in subjects performing prolonged and monotonous practical task simulations (Stave 1977, Thackray *et al.* 1977). Yet effort is not synonymous with stress. The specific function of effort is to maintain an optimal level of arousal for efficient performance in a given task. As originally defined (Selye 1976), stress is a syndrome of non-specific signs and symptoms reflecting an organism's struggle to maintain the homeostatic levels of vital parameters in the presence of physical factors that can disturb them. More recent 'psychological' definitions of stress (Lazarus 1966, McGrath 1970, Levine et al. 1978) extend that definition to the human anticipating the occurrence of a threat which may or may not disturb homeostasis, while at the same time appreciating his ability to cope with the threat using learned strategies. The non-specific reactions in this case depend upon the disparity between the anticipated threat and the anticipated ability to cope with it.

The combination of all factors that reduce arousal in repetitive work—repetitive task stimulation, constant environment, postural constraint—seems to be the physical stressor. Although diminished cerebral cortical arousal does not generally constitute a state of disturbed homeostasis, it becomes so when interpreted by the individual as a threat to successful performance. If the individual cannot continuously cope with this challenge he experiences anxiety. If he can cope only by means of great and unpleasant effort he experiences hostility. In the end, chronically high levels of effort, anxiety and hostility can be expected to produce psychological and peripheral physiological changes that constitute the full stress syndrome.

A dramatic description of stress occurring among workers in machine-paced, repetitive work is available in companion reports by Nerrell and Nilsson (1975). The workers were engaged in successive stages of sawmill production control. Each was required to make a precise visual discrimination, a decision and a contingent motor response within every successive two to three-second cycle of the work. Small mistakes due to inattention were immediately apparent to subsequent workers on the line, and larger errors brought the line to a halt. They reported experiencing great stress, anxiety, hostility, uncontrollable and unpredictable attentional lapses (blocking) and other mental abberations (Nilsson 1975). They also suffered to an extraordinary extent from psychosomatic disease (e.g. one-third overall, and as many as 60% in certain task categories were undergoing treatment for peptic ulcers or chronic gastritis at the time of the survey (Nerrell 1975).

To briefly summarize, highly repetitive work was thought to activate the habituation process causing de-arousal, and eventually performance failure. Phasic elevations in effort after recognized failures restore arousal but only temporarily, due to the tonic dependence of effort upon arousal. A major cause of stress in such tasks is, however, the individual's anticipation of the consequences of performance failures and his recognized inability to forestall them.

Numerous implications may be drawn from this theory. For example, rapid and invariant machine-paced, repetitive work is more stressful than more variable, but on the average, just as rapid self-paced work (Saito and Endo 1977). In the former case, it is impossible to prevent the coincidence of blocking episodes with attentional demands, and performance failures are more likely. Other implications regarding the workpace, introducing task or environment variety, allowing frequent rest pauses, etc., are fairly obvious, in accord with research findings and described in my previous paper (O'Hanlon 1981). However, some implications are more surprising and speculative. Within the normal range of perceptual-motor variety found in repetitive tasks, stress should increase with the attentional load required to make difficult perceptual discriminations and contingent decisions. This is because increasing perceptual difficulty raises the minimum arousal the worker must maintain in order to perform his task without failure. Greater effort is required to maintain higher arousal, and should effort diminish, the probability of performance failure is also greater.

5. A laboratory confirmation

The above notion contradicts a prevalent opinion that stress in repetitive work occurs because the human is 'underloaded' with respect to some ideal information-processing load (e.g. Welford 1968). Nonetheless, the prediction from the theory was recently confirmed in a study I shall now describe. Because the investigation was extensive and is reported fully elsewhere (Weber *et al.* 1980, Fussler-Pfohl *et al.* 1981), I will confine this description to a few salient points.

We asked male college students to perform four variations of a simulated ball-bearing packaging task. The subject's motoric task component was always the same, i.e., obtaining the separate units as each appeared from an automatic conveyor (two-second intervals) and placing it into a small container. When the container was full the subject stopped the conveyor just long enough to replace it with an empty container and then he continued as before. In the Simple Motor Condition (M) this was all the subject did. In the Counting Condition (C), the subject had to count the units into successive containers in lots of 60. In the Discrimination (D) Condition he had to discriminate between two ball-bearings (sizes 9 mm and 10 mm diameter) as these were presented in random order and equal proportions, placing each type in separate containers. Finally, in the Discrimination-Counting (DC) Condition he did both, by counting only the smaller ball-bearings in lots of 30 balls per container.

Among several measures related to work stress, we analysed integrated (30 s) EMG from over the trapezius and splenius capitis muscles at the back of the neck. EMG integrations were recorded on a strip-chart along with the raw data and a signal indicating whether the conveyor was on or off. We were therefore able to eliminate scores obtained during container replacement or at other times when the subjects engaged in gross, whole-body movements. As the subjects performed the various versions of the task, first for a short (14 min) period, and after a pause for interpolated tests, for a prolonged

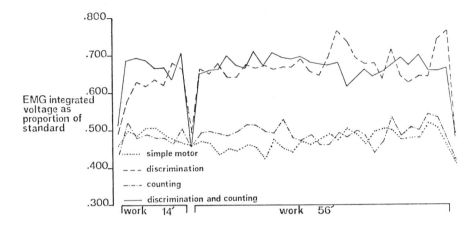

Figure 2. Task effects on EMG.

(56 min) period we noticed a profound difference in task effects on EMG. This is shown in figure 2.

EMG, and presumably isometric neck muscle tension, were considerably higher in conditions when the subjects had the perceptual task (D and DC) than in both of the others. But the counting task produced only slight elevation in tension relative to the simple motor task. Overall, the difference between EMG recorded under different task conditions was highly significant ($p \langle 0 \cdot 001$). Pair-comparisons between averages measured in four conditions confirmed the significance of the differences between either D or DC and C or M ($p \langle 0 \cdot 05$ to $p \langle 0 \cdot 001$), as well as the lack of significance between D and DC, and C and M.

Catecholamine excretion rates were also measured in this experiment. Overall adrenaline excretion rates exceeded resting values by about 50% and average rates varied between conditions in a manner analogous to EMG. However, the latter results were not significant. Another physiological measure of arousal (i.e., heart-rate variability) did vary significantly between conditions in accord with EMG. So also did the subject's assessments of relative tension or strain experienced in different conditions. We belatedly examined EMG data collected during a preliminary training session when the subjects had undertaken all tasks in a series of short trials. The differential effect of task condition on EMG was again readily apparent.

6. Conclusion

Certainly, this modest confirmation of a single prediction does not 'prove' the validity of the theory. But it does lead to the conclusion that further tests of this, and other, predictions are warranted.

References

ÅSTRAND, P. O., and RODAHL, K., 1977, *Textbook of Work Physiology* (New York: McGraw Hill).

BRANTON, P., 1970, A field study of repetitive manual work in relation to accidents at the work place. *International Journal of Production Research,* **8,** 93–107.

CAPLAN, R. D., COBB, S., FRENCH, J. R. P., Jr, VAN HARRISON, R., and PINNEAU, S. R., Jr, 1975, *Job demands and worker health* (Washington, DC: U.S. Department of Health, Education and Welfare).

FERGUSON, D., 1973, A study of occupational stress and health. *Ergonomics,* **16,** 649–663.

FUSSLER-PFOHL, C., WEBER, A., O'HANLON, J. F., BLAU, N., and GRANDJEAN, E., 1981, Tension musculaire de la nuque lors de travaux repetitifs. *La travaille humaine* (in the press).

GARDELL, B., 1971, Alienation and mental health in the modern industrial environment. In *Society, Stress and Disease: Volume I,* edited by L. Levi (New York: Oxford University Press).

GILBERTOVA, S., and GLIVICKY, V., 1967, Monotony at work (Czechoslovakian). *Studia Psychologia,* **9,** 232–240.

HAIDER, M., 1962, *Ermüdung, Beanspruchung und Leistung* (Vienna: Franz Deuticke).

HOCKEY, R., 1979, Stress and the cognitive components of skilled performance. In *Human Stress and Cognition: An Information Processing Approach,* edited by V. Hamilton and D. M. Warburten (New York: John Wiley).

JENNER, D. A., REYNOLDS, V., and HARRISON, G. A., 1980, Catecholamine excretion rates and occupation. *Ergonomics,* **23,** 237–246.

JOHANSSON, G., ARONSSON, G., and LINDSTRÖM, B. O., 1978, Social, psychological and neuroendocrine stress reactions in highly mechanised work. *Ergonomics,* **21,** 583–600.

KAGAN, A. R., and LEVI, L., 1974, Health and environment-psychosocial stimuli: A review. *Social Science and Medicine,* **8,** 225–241.

KAHNEMAN, D., 1973, *Attention and Effort* (Englewood Cliffs, N.J.: Prentice-Hall).

KASL, S. V., 1978, Epidemiological contributions to the study of work stress. In *Stress at Work,* edited by C. L. Cooper and R. Payne (Chichester, England: John Wiley).

KISHIDA, K., 1973, Temporal change of subsidiary behaviour in monotonous work, *Journal of Human Ergology,* **2,** 75–89.

KOGI, K., 1972, Repeated short indifference periods in industrial vigilance. *Journal of Human Ergology,* **1,** 111–121.

KORNHAUSER, A., 1965, *Mental Health of The Industrial Worker* (New York: John Wiley).

LAVILLE, A., and TEIGER, C., 1975, Santé mentale et conditions de travail. *Therapeutische Umschau,* **32,** 152–156.

LAZARUS, R. S., 1966, *Psychological Stress and the Coping Process* (New York: McGraw-Hill).

LEVINE, S., WEINBERG, J., and URSIN, H., 1978, Definition of the coping process and statement of the problem. In *Psychobiology of Stress: A Study of Coping Man,* edited by H. Ursin, E. Baade and S. Levine (New York: Academic Press).

MARTIN, E., ACKERMAN, U., UDRIS, I., and OERGERLI, K., 1980, *Monotonie in der Industrie* (Bern: Hans Huber Verlag).

McGRATH, J. E., 1970, A conceptual formulation for research on stress. In *Social and Psychological Factors in Stress,* edited by J. E. McGrath (New York: Holt, Rinehart and Winston).

NERELL, G., 1975, Medical complaints and findings in Swedish sawmill workers. In *Ergonomics in sawmills and woodworking industries,* edited by B. Thunell and B. Ager (Stockholm: National Board of Occupational Safety and Health), pp. 1–10.

NILSSON, C., 1975, Working conditions in the sawmill industry: a sociological approach based upon subjective data. In *Ergonomics in sawmills and wood-working industries,* edited by B. Thunell and B. Ager (Stockholm: National Board of Occupational Safety and Health), pp. 249–260.

O'HANLON, J. F., 1981, Boredom: Practical consequences and a theory. *Acta Psychologica* (in the press).

PRIBRAM, K. H., and McGUINNESS, D., 1975, Arousal, activation and effort in the control of attention. *Psychological Review,* **82,** 116–149.

ROMAN, P. H., and TRICE, H. M., 1972, Psychiatric impairment among 'middle Americans': surveys of work organizations. *Social Psychiatry,* **7,** 157–166.

SAITO, H., 1973, Studies on monotonous work (a summary) (Japanese). *Journal of Science of Labor,* **49,** 47–88.

SAITO, H. and ENDO, Y., 1977, *Monotonous Labor and Countermeasures for the Humanization of Labor* (Japanese), (Tokyo: Institute for Science of Labor).

SAITO, H., KISHIDA, K., ENDO, Y., and SAITO, M., 1972, Studies on bottle inspection task, I. Comparisons of different work control systems. *Journal of Science of Labor,* **48,** 475–532.

SAMOILOVA, A. J., 1971, Morbidity with temporary loss of working capacity of female workers engaged in monotonous work (Russian). *Sovetskia Zdravookhranenie,* **30,** 41–46.

SANDERS, A., 1979, Performance Theory (Chapter 8). In *Handbook of Psychonomics: Volume II,* edited by J. A. Michon, E. G. J. Eijkman and L. F. de Klerk (Amsterdam, New York: North Holland).

SELYE, H., 1976, *The Stress of Life* (New York: McGraw-Hill).

SIASSI, I., CROCETTI, G., and SPIRO, H. R., 1974, Loneliness and dissatisfaction in a blue collar population. *Archives of General Psychiatry,* **30,** 261–265.

STAVE, A. M., 1977, The effects of cockpit environment on long-term pilot performance. *Human Factors,* **19,** 503–514.

WEBER, A., FUSSLER, C., O'HANLON, J. F., GIERER, R., and GRANDJEAN, E., 1980, Psychophysiological Effects of Repetitive Tasks. *Ergonomics,* **23,** 1033–1046.

WELFORD, A. T., 1968, *Fundamentals of Skill* (London: Methuen).

Stress, pacing, and inspection

By C. G. Drury and B. G. Coury

State University of New York at Buffalo, Department of Industrial Engineering,
Amherst, New York, 14260, USA

Stress is shown to affect performance on inspection-like tasks and
inspection tasks are shown to be stress-producing. The speed stress of
pacing in a search-dominated inspection task is considered analytically,
where it is shown that unpaced performance is superior to paced
performance. A model of the inspector as a resource-limited information
processor is used to develop a methodology for the study of stress in
inspection.

1. Stress and inspection

The costs of inspection errors are rising. Product-liability suits and third-
party suits for work injuries are often a direct result of the failure of an
inspector to detect a fault in a product. Although the trend for inspection is
towards automation, such techniques are only suitable for easily quantified
faults such as electrical properties or discrete flaws. The problem of inspection
for less definable faults still remains a province of the human inspector. Thus,
as automated quality-control devices take over the basic detection functions of
inspection, the inspector is left with the complex decision-making tasks.

Despite an extensive history of research on inspection *performance*, little
has been said about the impact of the inspector's task on his/her wellbeing.
This was not so in the earlier papers, which tended to take a more holistic view
of inspection, with detailed and thoughtful case studies incorporating, at least
anecdotally, both the effects of the inspector on the system and the effects of
the system on the inspector.

There are two major components in the job of an inspector—search and
decision-making. Search is a sequential process, often visual scanning, which
ends with the location of a potential fault. Only when the search component
has found a flaw can the decision-making process come into play. Search is a
rather easy process for which to model performance (Drury 1977) but the
studies of inspection as a whole point to the decision-making as the important
contributor to stress.

There is, however, one overriding similarity between search and decision-
making: both are drastically affected by speed stress and pacing. All of
information theory from Hick (1952) onwards shows that the quality of a
decision is directly related to the time available to make the decision. Similarly,

search models back to Lamar (1946) show that the probability of success, i.e., performance, is directly affected by the time allowed for searching.

Even though speed stress affects both components of inspection *perform-ance*, it need not affect the stress inherent in the inspection task in the same way for each component. What is needed is the simultaneous measurement of performance *and* stress on an inspection task, plus a method for determining the locus of the effect of the stress. The remaining sections of this paper show how pacing affects search, how decision-making can be stressful and how a measure of the locus of the effect of stress on inspection might be achieved.

2. Pacing, inspection, and performance

Pacing by external means, as opposed to self-pacing, has been evaluated for inspection tasks a number of times with mixed results (Fox, 1977, Fox and Richardson 1970, Fox and Haslegrave 1969).

There are at least three aspects of pacing which can easily be confounded in inspection studies, leading to the above mixed results and to possible confusion in future stress research.

(1) *Speed of working*. In any inspection task the time allowed per item for inspection affects both types of error dramatically (Drury 1973, 1978). As more time is allowed per item, the inspector detects more faulty items and makes more false alarms.

(2) *Speed of movement*. If the items are moving past the inspector, visual acuity decreases with angular velocity. This may be important in conveyor-paced inspection (Buck 1975).

(3) *Rigidity of pacing*. Given the same *average* time per item, the pacing can be rigid or loose depending upon whether the inspector can allow a larger or smaller time for individual items. The *tolerance time* about the average time per item defines the rigidity.

A fact of life about inspection tasks is their extremely variable cycle time. For example, consider the simple inspection task of a systematic search for a single possible fault on an item. If it takes N fixations to search the item completely and perfectly then a randomly located fault will be found in a search time with a uniform distribution on the interval (0, Nt_o where t_o is the mean fixation time. The average time to detect a fault is $Nt_o/2$. If no fault is present, the time taken to search the whole item will be Nt_o, or exactly twice the mean search time. In practice, search times tend to be exponentially distributed rather than uniformly distributed with the probability of finding a fault in time t:

$$p(t) = \lambda e^{-\lambda t}$$

where $\lambda = 1/\bar{t}$, with $\bar{t}$ being the mean search time. For this distribution the standard deviation is equal to the mean. Hence, we have a situation which produces extremely variable search times. The stopping times for non-faulty items tend to be approximately normally distributed with a mean longer than twice the mean search time and a fairly low variance (e.g. Liuzzo and Drury 1980).

Clearly, any form of pacing of an inspection task must allow for this variety of performance times on each individual item. A reasonable question then becomes, how quickly should an inspector be paced, if at all, to achieve optimum performance? Recent research at SUNY at Buffalo has begun to answer this question, at least for inspection tasks where visual search is the major component. Morawski (1979) treated the time allowed per item as one of selecting a stopping time which balanced the costs of time spent searching against the value of finding a fault.

From the expressions for the expected values of the inspection outcomes, it can be proved that self-paced inspection *always* leads to a higher value than externally paced inspection.

Using the Morawski models it is possible to derive an optimum inspection time for a wide range of conditions where search is the major component of inspection. It is also possible to calculate whether inspection is even worthwhile; e.g., if the expected value is always negative, inspection is not a gainful activity.

From this discussion it is clear that pacing adversely affects search performance, but what of its impact on the decision-making component of inspection?

3. The stressful nature of complex inspection tasks

The effects of stressors on performance in complex tasks has been thoroughly and excellently reviewed elsewhere (Broadbent 1971, Mackie 1977, Hamilton and Warburton 1979).

There is evidence that inspection is a stressful activity. Krivohlavy *et al.* (1969), to cite one example, measured blink rate, accommodation and convergence time, and fine-reading ability of textile workers under conditions of low lighting levels. All measures showed reduced visual performance with time on shift.

Physiological indices have also been shown to be sensitive to stress·effects in inspection. A study of a continuous decision-making task (Thackery *et al.* 1974) showed a significant change in heart rate variability during a 40 min task but no change in mean heart rate. The type of job organization has also been shown to be significantly related to feelings of fatigue in a study by Kishida (1973). Jobs organized on fixed work-rate conveyors were found to adversely affect fatigue and subsidiary behaviour measures when compared to the indices for either variable-task conveyors or table-type work organization.

The experimental literature on stress also provides insights which are of interest to inspection research. The work in sustained attention and learning and memory (Hockey 1978, Broadbent 1978) has appealed to activation/arousal theory to account for the effects of stress. An intriguing conceptualization of human behaviour under stress has been suggested by Kahneman (1973). It is argued that the human organism has limited resources available to allocate to one process or another. When processes compete for limited resources, the organism allocates resources to those elements of the task deemed most relevant. This active selection process has been termed as 'mental effort' or 'selective attention', and the resource limit has been related to physiological arousal.

Hockey (1979) argues further that when the organism is faced with a stressful and arousing situation, a narrowing of attentional capacity occurs, with processing resources being allocated to fewer and fewer elements of the task. Performance decrements occur when arousal is at such a high level that the organism does not have enough resource capacity to attend to all the elements necessary to successfully complete the task. This notion of selectivity has been advocated by others (Baddeley 1972, Broadbent 1978) for a variety of stressors operating in complex decision-making and problem-solving tasks, and is also supported by physiological evidence (Warburton 1979).

The important point to be noted here is that the human is conceived as an *active* processor of information, seeking to optimize performance by allocating available resources in a manner consistent with the perceived and actual demands of the task. Stressors and their arousing capacity are seen in terms of their effect on resource capacity and allocation policy. Processing resources are allocated according to some internalized priority and value structure. Stressors, then, become an integral part of the overall demands of the task, and can have both positive and negative consequences for the human. This is a view consistent with our model of stress in inspection.

One needs to acknowledge, however, a distinction drawn by Norman and Bobrow (1975) between two types of processing: data-driven and resource-driven. Data-driven processing is limited by the quality of the input into the processing system. Resource-driven processing is limited by the amount of processing capacity which can be brought to bear on the task. Thus, stressors and their effects can be categorized in terms of their effect on the input to the system, and/or their impact on processing ability.

Lighting, for instance, degrades the quality of the input. Machine pacing may have a two-fold impact by either degrading the input or (through time stress) reducing the amount of time available for efficient decision-making.

4. A model of the inspector under stress

Inspection, then, can be viewed as a complex process, with the inspector assigned the role of assuring a constant performance output established by quality-control standards. The inspector must meet perceived and actual task demands by using resources he/she has available. Introduction of task stressors serves to increase the demands on the inspector. Depending on which process(es) or mechanism(s) are affected, the stressor's impact may be reflected in changes in performance. If the inspector is able to combat the stressor and its arousing effects by allocating more processing capacity, then any detrimental effects associated with a stressor would be reflected not in the inspector's performance, but in some other aspect of his/her behaviour or physiological response. Performance deteriorates when the inspector can no longer maintain the level of effort necessary to successfully complete the task.

Our approach to the study of stress and inspection seeks to identify the nature and characteristics of the effects of stress on performance, and reveal the locus of those stress effects. Much of the research cited in this paper illustrates how stressors affect performance. Few, however, reveal the impact of stressors on underlying processes. Consequently, many of the results are applicable only for stressors whose effects can be measured in terms of changes in total system output (i.e., overall performance measures).

5. A new technique for studying the inspector

Perhaps the major deterrent to investigating the complex aspects of inspection as well as the impact of various stressors on inspection performance has been the lack of an adequate technique and methodology for characterizing the decision-making process of the inspector. A researcher would like to be able to identify the inspector's internal representation—or his/her 'cognitive map', so to speak—of the decision process, particularly when the decision becomes more complex (i.e., the inspector must integrate information along a single or multiple dimensions) and the criteria used are not so obvious.

One technique which has considerable merit and is currently being employed in our laboratory is multidimensional scaling (Kruskal 1964). This is a process (based on least-squares regression) for representing objects geometrically in such a manner that the interpoint distances relate in some way to the similarities associated with the objects. The technique has been used extensively to assess concept learning and classification behaviour (Homa *et al.* 1973) and to track the evolving structure of a conceptual space (Homa *et al.* 1979). Not only does it provide a geometric configuration of a person's internalized representation of a task at any point in time, but it also provides a quantified index of the degree of structure.

The utility of this technique for studying inspector behaviour is clear.

Multidimensional scaling provides a means for: (1) revealing the nature and characteristics of the conceptual structure used by the inspector; (2) tracking the development of the inspector's conceptual structure during learning; and (3) monitoring the effects of stress on this conceptual structure.

Our approach to stress and inspection seeks to utilize multivariate measurement techniques to identify the nature and locus of the effects of different stressors on complex inspection tasks. Multidimensional scaling provides a distinct measure of a cognitive process. Standard performance measures (speed and accuracy) provide estimates of other behavioural processes such as psychomotor skill and information-processing ability. The physiological measures provide means for monitoring the organism's response.

One would expect, then, that different types of stressor would be reflected in the response measures as a function of that aspect of the system affected. For example, pacing may affect the inspector's search process, but not his/her decision ability. Thus, one might expect this stressor to be reflected in the speed and accuracy measures, but not the inspector's internal model. Changes in quality-control standards, however, should have an impact on the inspector's conceptualization of the task. How stressors with a distinct physiological response (such as heat stress) interact with these types of stressor is less obvious. The results of this effort will allow more accurate modelling of inspection behaviour, especially in complex tasks and situations.

References

BADDELEY, A. D., 1972, Selective attention and performance in dangerous environments. *British Journal of Psychology*, **63**, 537–546.

BROADBENT, D. D., 1971, *Decision and Stress* (London: Academic Press).

BROADBENT, D. E., 1978, The current state of noise research: a reply to Poulton. *Psychological Bulletin*, **85**, 1052–1067.

BUCK, J. R., 1975, Dynamic visual inspection: task factors, theory and economics. In *Human Reliability in Quality Control*, edited by C. G. Drury and J. G. Fox (London: Taylor & Francis).

DRURY, C. G., 1973, The effect of speed of working on industrial inspection accuracy. *Applied Ergonomics*, **4**, 2–7.

DRURY, C. G., 1977, *Stress and industrial inspection*, Final Report, DHEW (NIOSH), Cincinnati, Ohio, USA.

DRURY, C. G., 1978, Integrating human factors models into statistical quality control. *Human Factors*, **20**, 561–572.

FOX, J. G., 1977, Quality control of coins. In *Human Factors in Work, Design and Production*, edited by J. S. Weiner and H. S. Maule (London: Taylor & Francis).

FOX, J.G., and HASLEGRAVE, C. M., 1969, Industrial inspection efficiency and the probability of a defect occurring. *Ergonomics*, **12**, 173–721.

FOX, J. G., and RICHARDSON, S., 1970, *The complexity of the signal in visual inspection tasks.* Unpublished paper, University of Birmingham, England, Department of Engineering Production.

GREEN, D. M., and SWETS, J. A., 1966, *Signal Detection Theory and Psychophysics* (Hunington, N.Y.: Robert E. Krieger Publishing Co.).

HAMILTON, V., and WARBURTON, D. M., (editors), 1979, *Human Stress and Cognition* (New York: John Wiley).

HICK, W. E., 1952, On the rate of gain of information. *Quarterly Journal of Experimental Psychology*, **4**, 11–26.

HOCKEY, G. R. J., 1978, Effects of noise on human work efficiency. In *Handbook of Noise Assessment*, edited by D. N. May (New York: Van Nostrand-Reinhold).

HOCKEY, G. R. J., 1979, Stress and the cognitive components of skilled performance. In *Human Stress and Cognition*, edited by V. Hamilton and D. M. Warburton (New York: John Wiley).

HOMA, D., CROSS, J., CORNELL, D., GOLDMAN, D., and SCHWARTZ, S., 1973, Prototype abstraction and classification of new instances as a function of number of instances defining the prototype. *Journal of Experimental Psychology*, **101**, 116–122.

HOMA, D., RHOADS, D., and CHAMBLISS, D., 1979, Evolution of conceptual structure. *Journal of Experimental Psychology: Human Learning and Memory*, **5**, 11–23.

KAHNEMAN, D., 1973, *Attention and Effort* (Englewood Cliffs, NJ: Prentice Hall).

KISHIDA, K., 1973, Temporal change in subsidiary behaviour in monotonous work. *Journal of Human Ergology*, **2**, 75–89.

KRIVOHLAVY, J., KODAT, V., and CIZEK, P., 1969, Visual efficiency and fatigue during the afternoon shift. *Ergonomics*, **12**, 735–740.

KRUSKAL, J. B., 1964, Multidimensional scaling by optimized goodness of fit to a nonmetric hypothesis. *Psychometrika*, **29**, 1–27.

LAMAR, E. S., 1946, Visual Detection. In *Search and Screening*, edited by B. O. Koopman (OEG Report 56).

LIUZZO, J. G., and DRURY, C. G., 1980, An evaluation of blink inspection. *Human Factors*, **22**, 201–210.

MACKIE, R. R., (editor), 1977, *Vigilance* (New York: Plenum Press).

MORAWSKI, T. B., 1979, *Economic models of industrial inspection*. Unpublished master's thesis, Department of Industrial Engineering, State University of New York at Buffalo.

NORMAN, D. A., and BOBROW, D. G., 1975, On data limited and resource limited processes. *Cognitive Psychology*, **7**, 44–64.

THACKERAY, R. I., JONES, K. N., and TOUCHSTONE, R. M., 1974, Personality and physiological correlates of performance decrement on a monotonous task requiring sustained attention. *British Journal of Psychology*, **65**, 351–358.

WARBURTON, D. M., 1979, Physiological aspects of information processing and stress. In *Human Stress and Cognition*, edited by V. Hamilton and D. M. Warburton (New York: John Wiley).

A study of the working pattern of workers in buffered assembly systems

By H. J. WARNECKE, H. P. LENTES and H.-P. BARTENSCHLAGER

Fraunhofer-Institut für Arbeitswirtschaft und Organisation, F. R. Germany

The results of studies are presented in which the work patterns of workers
in buffered assembly lines were described as a function of the buffer level.
The necessary data was recorded in an assembly system for typewriters.
This assembly system was divided into sections with varying numbers of
parallel workplaces. These parallel workplaces were arranged as a shunt
system to a conveyor, so that workers carrying out the same assembly tasks
in one section could work independently of each other. Furthermore, the
assembly sections were separated by buffers of various sizes. The findings
regarding work patterns can be summarized as follows.

(1) The mean work rate is independent of the level of the downstream
buffer and is only influenced by the level of the upstream buffer in a few
cases.
(2) Allowance-time patterns in assembly sections with few parallel
workplaces are to some extent influenced by the levels of the upstream and
downstream buffers. Allowance-time patterns are largely independent of
buffer levels in the case of larger sections.

A simulation model is described with the aid of which the effects of
various work patterns on buffer levels were established, thus allowing
conclusions regarding the design of section buffers. The simulation runs
were carried out with data for an assembly system with five sections using
various work patterns established during the studies. The findings allowed
the following conclusions regarding the dimensioning.

(1) In the case of two small assembly sections a relatively small buffer is
sufficient in order to ensure adequate buffering of the sections if work
behaviour is independent of buffer levels.
(2) Buffers between two large sections must have a considerably larger
capacity in order to ensure adequate buffering where work patterns are
independent of buffer levels.

1. Introduction

One of the most significant factors of non-buffered assembly lines in the
past was the dependence of the execution of assembly operations on other
workstations and on the conveyor. When planning new assembly systems this
dependence can be reduced or completely eliminated by means of appropriately dimensioned buffers. The possibilities offered to workers by buffers are,
for example:

(1) freedom to determine own workrate;
(2) freedom in taking breaks; and
(3) freedom to undertake indirect activities.

In addition to these worker-related aims, there are also technical and economic aims to be considered when designing buffers. Until recently, these technical and economic factors were the main feature of the procedures for designing buffers (Lentes 1978). Comprehensive studies (Bartenschlager *et al.* 1979) showed, however, that this approach neither allowed appropriate consideration to be given to worker-related aims of buffering during the planning phase, nor successful realization during implementation. In addition to the technical and the economic aspects it is important that:

(1) the planner has a knowledge of working patterns of employees in buffered assembly systems; and
(2) the employees know the function of buffers.

The first findings of studies on these problems are reported in Bartenschlager *et al.* (1979). These findings and the findings of this paper are the results of work carried out at an office machine manufacturer as part of the government-funded programme 'humanization of the work environment'.

The main aims of this project are:

(1) the creation of larger and more complete work contents; and
(2) greater freedom of action for the individual worker by reducing dependence of workers on each other and on the speed of the conveyor.

2. Description of the area studied

The studies were carried out in a work system for the assembly of electric typewriters. The above aims would be achieved by implementing the following measures:

(1) division of assembly into three subsystems; the assembly of the body, of the carriage and completion, including adjusting and checking operations;
(2) grouping of individual operations to form larger and complete work contents;
(3) execution of these work contents in assembly sections which shunt-type parallel workstations; and
(4) the separation of the individual assembly sections by buffers.

Figure 1 shows the layout of one of these assembly sections with six parallel workplaces. The same work content is to be carried out at each of these workplaces. In this case the time per workplace per product is six minutes, because the average output interval of an assembly section is given as approximately one minute.

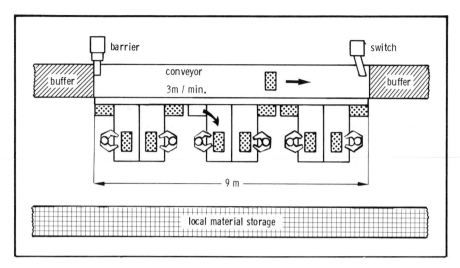

Figure 1. Arrangement of parallel workplaces in one assembly section.

The arrangement of the workplaces next to the conveyor allows the employees to work independently. After completing work on a typewriter, the worker pushes it onto the conveyor. The position of the typewriter on the conveyor indicates whether it has been completed or not. This and the arrange-

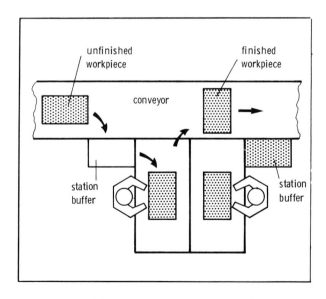

Figure 2. Work flow at workplaces with station buffers.

ment of the station buffer, from which the worker pulls the next typewriter, is shown in figure 2. Without this station buffer with the capacity of one typewriter, an employee would have to wait for a machine to arrive on the conveyor, whereas now he merely pulls the machine from the station buffer and starts work. The completed machine is transported on the continuously moving conveyor to the next section buffer and operates a switch as it enters. This switch withdraws the barrier at the beginning of the section and allows another unfinished machine to enter the section. It is transported to the worker who put the last completed machine back on the conveyor, and he pulls it into his station buffer.

This organization of the work flow ensures that the workers in any one assembly section can work largely independently of each other and of subsequent assembly sections. Furthermore, there is no dependence on the conveyor if it is ensured that the time required for transport to a station is considerably less than the average work time, in which case workers do not have to wait for the next.machine even after working at an above-average speed.

3. Problem

One of the problems in planning the above work system is the dimensioning of the buffers between the individual assembly sections. As this project involved redesigning an existing assembly system and there was relatively little experience of buffer dimensioning for such assembly systems, an attempt was made to improve an initial solution step by step in order to achieve the best possible final solution. The number of steps and as a result the number of changes was minimized by simulating the entire work flow with a model. This gave some leads for the design of the system.

In order to make the model as realistic as possible it was necessary to have data on the behaviour of the system. In addition to deterministic variables including:

the number of assembly sections;
the number of workplaces per section; and
the length of buffers;

the random variables describing the work patterns of the workers are significant.
These include:

average work times;
the dispersion of work times; and
the frequency and duration of allowance times..

In order to describe the system behaviour it is necessary to know not only the values of these variables, but also whether these variables are inter-dependent and whether the state of the system at any one time affects these variables. It is particularly important to know whether work and allowance time patterns are affected by the buffer levels.

In order to clarify this situation, the work and allowance time patterns of workers in various assembly sections were studied and the resulting data were statistically evaluated.

4. Approach

The study was carried out after reorganization of typewriter assembly. During reorganization the measures indicated above were implemented. The buffers were dimensioned in such a way as to allow each worker to leave his workplace once a day for up to 15 minutes 'at one time', without affecting neighbouring assembly sections.

The body-assembly subsystem was chosen as the object of the study. Here, the individual sections have two to nine parallel workplaces, which at an output interval of one minute means work contents per assembly section lasting between two and nine minutes. The buffers between the sections had a capacity of 10 to 35 machines.

In order to ensure that the work pattern findings were as comprehensive as possible, the assembly sections studied were selected to include sections with many and sections with few workplaces. A further criterion for selection was the inclusion of sections between long buffers and sections between short buffers. (The method for data collection is reported in Bartenschlager *et al.* 1979).

The data resulting from these studies were then statistically evaluated. In addition to univariate procedures for the analysis of distributions, the cycle and allowances times were subjected to correlation and regression analysis in order to show possible correlations between work patterns and buffer levels.

5. Findings

5.1. *Correlation between cycle times and buffer levels*

As described in Bartenschlager *et al.* (1979), the statistical evaluation of the cycle times and levels of the upstream buffers showed significant correlations for some workplaces. In these cases, the average cycle time was slightly shorter when the upstream buffer was full than when it was empty.

In the cases where the workers sat with their backs to the buffer or if the workplaces in large assembly sections were at considerable distance from the end of the buffer, the correlations were not significant.

Regardless of the size of the assembly section and the arrangement of the workplaces, no significant effect of the downstream buffer on the cycle times of the workplaces was established.

Thus, the work pattern of the workers can be summarized as follows. If workers sit in the vicinity of the upstream buffer and can see the level, they tend to work faster if the buffer is full than if it is empty. The level of the downstream buffer has no effect on the speed at which work is carried out.

5.2. *Correlation between the duration of allowance times and buffer levels*

The allowance times which could be determined by the workers themselves were chosen to study the correlation between allowance times and buffer levels. Thus, allowance times resulting from technical, organizational or process-related problems were disregarded. Similarly, the times for tidying and cleaning the workplace were not included in the analysis, as these activities must take place at a certain time of day.

The times studied included the following:

(1) personal allowance times at or in the vicinity of the workplaces;
(2) personal or work-related allowance times away from the workplace;
(3) time for replenishing material containers; and
(4) time for discussion with setters and foremen.

In the statistical analysis the relationship between the duration of an allowance time and the buffer level at the beginning of the allowance time was

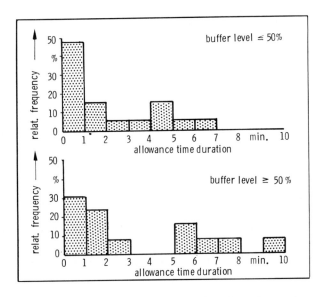

Figure 3. Relative frequency of allowance-time durations for various downstream buffer levels.

examined. The evaluation showed not only significant connections between the duration of allowance times and the level of the upstream buffers as in the case of cycle times, but more significantly, influences of the level of the downstream buffer on allowance times.

This is shown in figure 3. The section used as an example consists of six parallel workplaces. In each case the two workers sit opposite each other. As a result, three of the six workers can see the level of the downstream buffer.

The upper section of figure 3 shows the relative frequency of the duration of allowance time recorded when the downstream buffer was less than half-full. In this case, the large majority of allowance times was less than two minutes.

Only 10% of allowance times lasted longer than five minutes. The lower section of the figure shows the relative frequency of allowance times recorded when the buffer level was high.

Here too, the duration of most allowance times is short. The proportion of longer allowance times—approximately 38% of the allowance times were longer than five minutes—is, however, considerably higher when the buffer level is high.

The reverse was observed in the case of the upstream buffer. Here there was a tendency for the allowance time to be shorter if the upstream buffer was comparatively full than if it was comparatively empty.

5.3. *Correlation between the frequency of allowance time and buffer levels*

The evaluation of the data showed that the buffer levels not only influenced the duration of allowance time, but also had some influence—if only slight —on the frequency of allowance times.

In some sections more use was made of allowance times if the upstream buffer was at a low level.

The level of the downstream buffers had the reverse effect on the frequency of allowance times. Workers made greater use of allowance times if the level of the downstream buffer was comparatively high than if it was low.

However, the observed correlation only occurred in the case of assembly sections with few workplaces. In the case of larger sections, no influence of the buffer levels on the allowance time patterns was established.

Thus, the allowance-time patterns of the workers can be summarized as follows: in sections with few workplaces, the workers tend to make less use of allowance time if the upstream buffer is full and the downstream buffer is empty. Thus, in sections with a small number of workplaces, the allowance time patterns are to some extent adjusted to the respective buffer levels.

In larger sections, in which one worker making use of his allowance time has only a slight effect on the total output of the section, the allowance time behaviour is independent of the respective buffer levels.

6. Simulation of work processes

6.1. *Description of the simulation model*
The random variables of the simulation model were based on the following premises regarding distribution:

(1) cycle times are normally distributed with a variability ranging from 50% to 200% of the mean;

(2) the durations of allowance times are distributed exponentially;

(3) the intervals between two allowance times are distributed exponentially. One restriction was imposed, namely that two allowance times must be separated by at least one cycle;

(4) the cycle and allowance times of the individual workers do not affect those of other workers;

Statistical tests showed that the distributions resulting from the above premises showed good agreement with observations.

In order to represent the dependence of cycle and allowance times on buffer levels in the model, the following solution was chosen. The mean cycle time is, for example, 10% larger when the buffer is empty, and 10% lower when it is full, than the mean cycle time with a half-full buffer. By varying these percentages, the degree of influence of the buffer level on the cycle times can be varied. If a percentage of zero is chosen, this means that the mean for all buffer levels is the same and thus independent of the buffer.

In the same way as described here for the mean value of cycle-time distribution, the mean allowance times and intervals can also be varied as a function of the respective buffer level by means of such percentages.

The simulation model briefly described above and written in FORTRAN IV was used to study the effects of various work patterns on buffer levels and thus on the required buffer capacities, whereby the work patterns described above were taken into account.

6.2. *Execution of simulation runs*
In order to establish the effects of various work patterns, simulation runs were carried out which corresponded to a working period of 500 minutes each. An assembly system with five assembly sections was simulated. The first two sections had four parallel workplaces each, the next two eight workplaces each and the last section four workplaces. This allowed the effects of buffer levels for the following constellations to be studied:

(1) two buffered sections with four workplaces each;

(2) two buffered sections with eight workplaces each; and

(3) two buffered sections with four and eight workplaces respectively.

Each buffer was given a capacity of 20 workpieces and an initial level of ten workpieces. This was done in order to ensure that the work system would quickly reach an 'operational' condition during simulation.

The parameters for cycle times were selected in such a way that the mean output interval of a fully-manned section (i.e. where no worker was missing as a result of an allowance time) was one minute. Thus, the mean cycle time for a worker in a section with four workplaces was four minutes and the mean cycle time for a worker in a section with eight places was eight minutes.

The standard deviation was set at 20% of the mean in each case. This corresponds to the findings of the study, which showed variation coefficients between 0·1 and 0·3, which corresponds to an average of approximately 0·2.

Evaluation of the collected data showed a mean allowance time duration of approximately two minutes and a mean allowance time interval of approximately 60 minutes. These values were used for all workplaces during the simulation runs.

6.3. *Findings*

The simulation findings for the following two cases are to be described as examples of the effects of various work patterns established by means of simulation:

(1) work patterns are independent of buffer levels;

(2) work patterns are influenced by the levels both of the upstream and of the downstream buffers.

As one of the main aims of the study was to establish guidelines on the

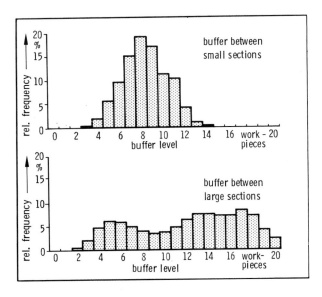

Figure 4. Frequency distribution of buffer levels where work patterns are independent of buffer levels, as established by simulation.

dimensioning of section buffers, the levels of the buffers and the frequency distribution of these levels were monitored and analysed for the above cases.

Figure 4 shows the results obtained for frequency distribution for the first case, in which the work patterns, i.e. the distribution of cycle and allowance times, were assumed to be independent of buffer levels. The values for relative frequency of the respective levels are mean values resulting from several simulation runs conducted with different initial random numbers.

The upper section of figure 4 shows the levels for the buffer located between two sections with four workplaces each. The level lies between relatively narrow limits, whereby the greatest frequency is in a range from six to 11 workpieces. Thus, the capacity of this buffer, namely 20 workpieces, is not fully used. However, this does not apply for a buffer of the same capacity located between two sections with eight workplaces each. This is shown by figure 4. The buffer levels are equally distributed thoughout the entire range, showing no preferred level.

In the second case examined here, the work patterns are influenced by the level of both the upstream and the downstream buffers. The influences of these buffers were programmed as described in the simulation model. For the mean cycle time a variation of ± 10%, for the mean allowance time of ± 20% was chosen. The variation was such that the mean output of a section increased when the upstream buffer was full and decreased when it was empty, and vice-versa for the downstream buffer.

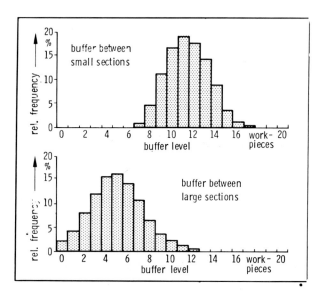

Figure 5. Frequency distribution of buffer levels where work patterns are influenced by the level both of the upstream and downstream buffers, as established by simulation.

The distribution of buffer levels obtained using this simulation model is shown in figure 5.

The distribution of buffer levels for the buffer between two sections with four workplaces each remains basically unchanged compared to the first case. The dispersion range is ten typewriters. The frequency distribution has, however, shifted in the direction of higher buffer levels, so that the maximum buffer level has increased by three.

The buffer level distribution for the buffer between two sections with eight workplaces each shows considerable changes compared to the corresponding diagram for the first case. Whereas the levels for uninfluenced work patterns were evenly distributed throughout the whole range, there is now an accumulation on the left side of the range.

6.4. *Conclusions*

The simulation results described briefly here allow conclusions to be drawn regarding the design of buffers.

The distribution of buffer levels for buffers between two small assembly sections is almost independent of the workers' work patterns. At a mean output interval per section of one minute, the dispersion of buffer contents can be largely accommodated by a capacity of ten typewriters.

If a buffer is located between two large assembly sections, the dispersion range for buffer levels for independent work patterns is considerably higher than for work patterns affected by buffer levels. In the first case, the buffers must be of a correspondingly large capacity.

From the economic point of view, it would be expedient in large work sections if the workers were to adjust their workrate to differing buffer levels. However, this would impose a restraint on the worker, so that the buffer would not be able to fulfill one of its most important functions, namely of allowing the worker greater freedom of action. Thus, the aim should be to provide the buffer capacity required for independent work patterns. If this requirement cannot be completely fulfilled, as a result of limited space or the investment involved, two things should be done. On the one hand, the arrangement of the workplaces and the buffers should be such that the workers can easily see the buffer levels. On the other hand, the workers should be taught the functions of buffers. These measures allow employees on assembly lines with buffers to increase the effectiveness of buffers by adopting appropriate work patterns.

This paper represents a first approach to the development of rules for the dimensioning of buffers when taking worker-related aims into account. In the future it is necessary to add to these design guidelines, paying particular attention to sociological aspects.

References

BARTENSCHLAGER, H. P., KOHL, W. and WARNECKE, H. J., 1979, Methoden zur Untersuchung des Arbeitsverhaltens an gepufferten Arbeitsplätzen. *Zeitschrift für Arbeitswissenschaft 33 (5NF)*, No. 1, S. 7–14.

LENTES, H. P., 1978, Auslegung von Montagelinien mit Hilfe der Simulation. *VDI-Berichte No. 323*, S. 19–28.

Section 6. Impact of machine-paced work on stress

Symptom levels in assembly-line workers[†]

By D. E. Broadbent and D. Gath

Departments of Psychology and Psychiatry, Oxford University

This paper describes the results of surveys of mental health in two plants undertaking motor-car manufacture. In both cases, repetitive workers showed lower job satisfaction than non-repetitive ones; but not necessarily poorer health. Those who were paced by an assembly line, however, showed higher anxiety than those who were not; with no difference in job satisfaction. Cycle time, over the range found in these plants, bore no relation to health or to satisfaction.

1. Background

Modern methods of mass production are often accused of causing harm to the health of the worker. They require people to repeat very simple actions over and over again, and this itself allows little exercise of skill or initiative; in some cases the cycle of operations is very short before it repeats, and in some cases the timing of the sequence is under the control of machines rather than the workers themselves. These three features of work (repetition, short cycle time, and pacing) are often lumped together and condemned out of hand; and yet a job can quite well have one of them without the others. From a practical point of view, it would be valuable to know which features of a job were the most undesirable and should be altered, and which features were unimportant.

In any case, there is little hard evidence that health is poor in jobs of this kind because of the work itself, as distinct from general social factors. People in unskilled repetitive work do frequently have a general feeling of demoralization or malaise, but that might mean only that they are discontented with their general social position or possibly that people with personal problems drift into unskilled jobs. No research seems to have shown connections between particular features of the job and particular disorders of the person.

Accordingly, the Department of Employment and the Employment Medical Advisory Service commissioned from the Medical Research Council further studies on this topic. The following account is of research carried out by the Departments of Psychiatry and of Experimental Psychology, Oxford University, under this commission.

[†] A similar version of this paper has appeared in the Department of Employment Gazette.

2. Preliminaries

Medical records from a large motorcar manufacturing plant showed that patients from the assembly line contained a higher proportion of cases diagnosed as 'anxious' than did patients from elsewhere in the plant. This might merely mean that the doctors making the diagnosis were unconsciously biased by their knowledge of the man's work; but it suggested the use of standardized questions which could be given by an interviewer to workers having different kinds of job, and which would discriminate different forms of ill-health. As a starting point, some questions were taken from the Middlesex Hospital Questionnaire. These questions had already been shown to give answers which were systematically different in healthy people on the one hand, and in patients on the other. Some alterations were needed to make them acceptable in an industrial situation, but the revised questions still gave different answers in people taken from the general population and from a psychiatric clinic. When the revised questions were given to a small sample of laundry workers, the people with the greatest degree of pacing tended to have the highest scores on anxiety, though not on other kinds of symptom. It therefore seemed worthwhile to give a standard interview, including the MHQ questions, to larger groups of workers with known degrees of pacing, cycle time, and repetition.

3. The interview

In its final form, each interview started with a number of questions about the physical details of the person's job. These asked about the extent of repetition, pacing, and cycle time as these factors were the main interest of the research; but also about the physical effort involved, the degree of concentration or diffusion of attention, the risk of possible error, the extent to which the work could be seen as fitting into a general purpose, and so on. Apart from the work itself, questions were also asked about the environment, about the distance travelled to work, and other background factors such as family responsibilities. The person's main motives in working, and his degree of dissatisfaction with his present job, were also obtained.

After the job questions, there followed 28 questions based on the MHQ. These provided measures of anxiety (feelings of tension and worry), depression (lethargy and inability to make an effort), somatic symptoms (stomach upsets, giddiness, and similar bodily sensations known to increase when mental health is bad), and obsessional problems. The latter require further comment; the original set of questions asked about difficulties in the control of thought and action which are often seen in psychiatric practice, such as continual thinking about unwanted matters, or excessive punctilious washing. However, in this research it was found that the questions fell into two groups, and people who scored highly on one group had no tendency to score highly on the other. There are other investigations which have found similar results, so it seems wise to divide the obsessional questions into two sets. One of these concerns recent failures of control ('unwanted thoughts') and will be called obsessional symptoms. The other questions concerned long-lasting habits of thought (conscientiousness or perfectionism) and will be called obsessional personality.

The third section of the interview consisted of questions about cognitive failures, that is, minor slips of memory or attention in everyday life. They can be divided into groups according to their tendency to occur in the same individuals; these groups are known simply as A, B, C, and D.

The fourth and last section of the interview consisted of questions about the use of health services, such as visits to the doctor, taking of medicines both prescribed and unprescribed; and also a series of questions about the use of leisure.

The interview was always given by one of three women interviewers, each aged within the range of the men being studied, and having some professional qualifications in health care. Individual interviewers gave slightly different results, but none which affect the following conclusions. In selecting individual people for interview, the original approach was to start from nominal rolls of each work group. A proportion of names equally spaced through the list was then picked to give a sample of desired size, and as far as possible these names were then the first to be asked if they were willing to take part. If they were not, the next name on the list was then approached, and so on. The actual approaches were made by the man's own supervisory staff or by the appropriate shop stewards (who had of course given general agreement to the study at an earlier stage). This method of approach had the advantage of making it clear to the man that the research had full backing of union and management; and indeed was probably essential. It should be noted, however, that it may have resulted in uncontrolled differences in the way the proposal was put to the men. There were also major operational problems in that some men were hard to spare from their work on certain days. No satisfactory measure of the true response rate is therefore available, and from a technical point of view this is probably the main weakness of the study. We did not find, however, any difference between the men willing to be interviewed and the general population of the work group, in anything which might affect the following results.

4. Plan of main investigations

Because of the problem of social factors, mentioned above, it was essential to compare people whose jobs differed but who were as similar as possible in other factors such as pay, home neighbourhood, and so on; the role of many of the questions in the interview was merely to confirm that no unsuspected difference had crept into the comparison. All the workers studied were therefore men, employed by the same company engaged in the manufacture of motor vehicles. Two plants were examined, Plant A, producing bodies 'in white', that is, before the stage of final painting, and Plant B, conducting the final assembly of cars by fitting together the painted body, engine, transmissions, etc., to complete the car. Comparisons were made within each plant since each contained relatively paced and relatively unpaced work; thus, the comparisons should eliminate local factors of plant morale, quality of environment, and so on. As is now common in the motor industry, the policy of this company is to pay all direct-production workers equally, with no incentive payments based on individual output, so this factor also is eliminated from comparisons between production workers in the same plant. The skilled toolroom workers in Plant A were paid very slightly *less* than production workers.

In Plant A, the original hope was to compare men working on assembly itself with those working presses, the notion being that the latter would be unpaced. This approach turned out to be misguided, because in pressing large metal components such as car doors, the door is moved from press to press without interruption and a whole series of operations performed upon it. A worker on one press in the middle of the series must therefore complete his task on each item before the next arrives from the previous press; he regards himself as paced even though there is no visible conveyor belt. In addition, press workers in this plant changed their jobs from day to day depending on the particular needs of production. Conversely, the group of assembly workers was not homogeneous; some were indeed required to complete a task (e.g., welding a door-post into the body) while the bodies moved past on a conveyor, and this task had to be completed within a rigidly defined time. Other workers, however, could perform similar tasks very much at their own speed, because they were producing a 'bank' of assemblies for some later stage of the process to use. A man welding pieces of metal together to form door-posts could, by going slightly fast for a period, get ahead of himself and relax for a period without any serious consequence.

Ultimately, then, 19 such unpaced assembly non-line men (AN) were compared with 23 paced assembly-line workers (AL), and 48 press workers were regarded as semi-paced or intermediate (PR).

Plant A also employed a number of skilled toolroom workers, whose jobs were by most standards non-repetitive. Their work was to produce new or replacement dies and jigs for subsequent use in production; for an individual fitter each task might take several weeks and involve a number of different processes. Machinists, as opposed to fitters, stayed by a single machine capable of, for example, cutting a final version in metal from an original pattern made up in less durable material; again each task would be different. There were two toolrooms, differing in age and therefore in quality of environment; in Toolroom 1, 38 fitters and 27 machinists were interviewed, and in Toolroom 2, 27 fitters and 23 machinists. It is worth noting that nine of the fitters in Toolroom 1 worked on jigs rather than dies, and appeared rather different from those on other types of work; they were therefore kept separate in some of the analyses.

In Plant B, it did not prove possible to make a comparison of paced and unpaced workers within the assembly process itself. Although a few men engaged on assembly had a bank of work, they were coupled closely to the flow of cars and could not win breathing space for themselves by building up the size of the bank. The actual process of assembly did not employ a mechanical conveyor; rather, the line of cars was pushed on by hand from one position to the next as each operation was completed. However, as there was no space between cars, any man who failed to complete his task by the time those before and after him had finished would hold up the entire flow. This produced considerable social pressure to keep up, as in the press workers of Plant A. These men could be regarded therefore as paced; 45 of them were interviewed and can be called repetitive paced assemblers (RPA).

For an unpaced comparison, it was necessary to go outside assembly to a group of 11 men engaged in rectification, tuning, etc., of cars requiring major attention. The speed of this work was very much under the control of the man himself, though the job tended to change from day to day, and the group is therefore known as non-repetitive unpaced (NUB).

There were two intermediate groups in Plant B, who could not be described confidently as paced or as unpaced. One group of 12 was made up of relief workers and rectifiers working on the loss; they were sometimes paced and sometimes not, depending on the job they were doing, and are described as line-reliefs and rectifiers (LRR). The second doubtful group, also of 12 men, carried out rather mixed but repetitive jobs such as paint-spraying. They were not tied to the line itself but were unable to work far enough ahead to make a breathing space for themselves; they are termed repetitive mixed (RM).

Thus, a grand total of 285 men were interviewed, although the main interest centres on the 68 who were unequivocally paced and the 30 who were reasonably comparable but unpaced.

5. Results

It is easiest to explain the results by going first through those for Plant A, to illustrate the findings, and then showing that findings are similar for Plant B. In each case, results must of course be larger than the chance differences between people if they are to be regarded as trustworthy.

5.1. *Plant A*.

Firstly, the effect of *repetitive work as such*: we can compare the 115 skilled

Table 1. Plant A: differences between repetitive and non-repetitive workers in satisfaction, between paced and unpaced in anxiety.

	Dissatisfaction (Maximum 3)	Anxiety (Maximum 14)
Non-repetitive		
Toolroom workers	1·18	2·10
Repetitive		
Paced-assembly workers	1·56	3·13
Semi-paced press workers	1·50	2·5
Unpaced assemblers	1·53	1·58
Statistical significance	Repetitive different from non-repetitive	Repetitive not different from average of non-repetitive
	Pacing no effect	Paced different from unpaced

workers and the 90 production workers, and the latter clearly feel less satisfied with their jobs, as shown by the question 'How much of the time do you feel fed up?' The difference is comfortably larger than the chance differences between people. It is chiefly due to boredom; whereas the non-repetitive workers attribute any dissatisfaction largely to frustration and difficulties arising in getting things done. On the other hand, there is no evidence that repetitive workers are any less healthy overall than the skilled workers are; the differences in anxiety, somatic symptoms, obsessional symptoms, and depression are all less than the chance differences between people. So repetition seems to go with being unhappy, but not necessarily being unhealthy.

Secondly, the *differences of paced and unpaced repetitive work*: there is a marked contrast in the results in the second comparison. The paced and unpaced assemblers, AL and AN, appear equally dissatisfied; any difference is much smaller than the chance variation. But the two groups do differ in symptoms of anxiety, the paced workers showing a higher level. The semi-paced press workers are intermediate in anxiety, so paced work seems to go with this kind of symptom, even in a case where there is no difference in happiness with the job.

Thirdly, the effect of *cycle time in repetitive work*: this is a surprising negative result; there is no evidence that short cycle times are associated either with dissatisfaction or with ill-health, when they are compared with repetitive jobs with cycle times of up to half-an-hour or so. The press workers, whose cycle time was typically under a minute, were not particularly dissatisfied or

Table 2. Plant B: confirmatory differences.

	Dissatisfaction	Anxiety
Non-repetitive unpaced rectifiers, etc.	0·54	1·0
Repetitive paced assemblers	1·58	2·31
Statistical significance	Difference clearly established	Difference borderline one-tail significant

high in symptoms, compared with the paced assemblers with longer cycle times. Within the two groups of assemblers, the people with the shortest cycles gave only chance differences from those with the longest. If anything, they were healthier and happier, but this was probably only a chance difference.

For completeness, perhaps it should be mentioned that Plant A showed a few tentative results which were not confirmed in Plant B. For instance, the paced assemblers AL had more obsessional symptoms than the unpaced AN; but this effect disappeared in Plant B, so it was probably due to some local factor in Plant A. The two groups also showed quite an ominous difference in depression, even though it was still too small to be regarded as more than chance; but this effect also vanished in Plant B.

5.2. *Plant B.*

On the main findings, however, the agreement was good. Let us consider *pacing.* The paced assemblers RPAS, though less anxious than those in Plant A, still showed a higher level of anxiety than the unpaced comparison NUB. The difference might not indeed be proof of a relation between pacing and anxiety if it stood alone, but is acceptable as a confirmation of the earlier finding. (In a statistician's terms, it is 'one-tail significant'.)

Like the press workers of Plant A, the partly paced groups of Plant B, LRR and RM, showed quite high levels of anxiety though not as clearly different from chance as the RPAS.

So far as *job satisfaction* went, the unpaced and non-repetitive group NUB was the least fed up we have examined, and clearly happier than the assemblers who were as dissatisfied as the assemblers from Plant A. Once again, the assemblers complained of boredom while the comparison group, if fed up, attributed it to frustration. RM, whose work remained the same from day to day, resembled the assemblers in dissatisfaction while LRR was intermediate; in fact the latter group included some who did the same job each day and were as dissatisfied as the assemblers. The rest changed jobs each day and were about as happy as NUB. There does still seem, therefore, to be a relation between repetitive work and dissatisfaction.

As regards *cycle time*, once again there was no sign of any relation other than chance, either to symptoms or to dissatisfaction. It should be emphasized that even the longest cycle time was only 12 minutes, so it would be dangerous to draw major conclusions. However, we still have no positive evidence that short cycle times are any different from long, given that the job is repetitive.

5.3. *Relationship to personality*

The points already made are the main findings; clearly there are a large number of other relationships in the data which will be worth further analysis. As one example, remember the score of obsessional personality, mentioned earlier. Suppose we divide up the repetitive workers into those with high and low scores on this quality: the high scorers are men who are rather meticulous, conscientious, and precise. Let us call them meticulous, and the remainder relaxed. The meticulous tend, for example, to say that they look for 'satisfaction' in their job, whereas the relaxed are more likely to say 'money'.

The relationship between pacing and anxiety, mentioned already, is mainly due to the meticulous workers. Although we are now of course dealing with smaller numbers, in Plant A the difference in anxiety between the paced and unpaced meticulous is safely bigger than the chance variation; and in Plant B it is 'one-tail significant' again. For the relaxed, however, the differences are so small that they may well be due to chance. The fact that the paced workers in Plant B gave rather lower anxiety than those in Plant A turns out to be because they include rather fewer meticulous personalities.

It seems, therefore, that there is a kind of personality who is particularly unsuited to paced work; and it certainly seems reasonable that a man who likes to check his work may be especially likely to get anxious if he has no control over the speed of his operations. On the other hand, the relaxed personalities

Table 3. Relation of pacing to personality.

	Anxiety scores	
	Meticulous personalities	Relaxed personalities
Plant A		
Paced assemblers	3·31	2·71
Unpaced assemblers	1·0	2·0
Plant B		
Paced assemblers	2·85	2·06
Non-repetitive unpaced	1·1	Too few to score
Statistical significance	Plant A: Paced different from unpaced Plant B: One-tail significant	No significant difference

get just as fed up with their jobs as the meticulous, and there is no difference between the two kinds of men in the effect of repetitive work on job satisfaction. It is only when we are considering pacing, and its relation to anxiety, that it matters whether a person is meticulous or relaxed.

6. Meaning and limitations of these results

These findings do seem to be an advance on the state of knowledge when the work was started. By making comparisons between people in the same workplace, we have improved the evidence that differences in health and satisfaction are linked to the job itself and not only to social factors. There is also a strong suggestion that there is a split between the effects of repetition (which mostly goes with discontent) and those of pacing or lack of control over speed of work (which mostly goes with anxiety). On the other hand, the common suspicion of short cycle times gets no backing from this work.

Does the higher level of anxiety matter? Perhaps the best answer to this is to compare the levels found in our plants with those in a group of patients who have had to seek psychiatric help. Their scores of anxiety are higher on average than those in the car plants; but there is of course some overlap between the lowest scores in the patients and the highest in the working population. The best discrimination occurs if we take a score of five or above as a sign that the man is in some difficulty. That score marks a dividing line, above which the proportion of patients with such a score is higher than the proportion of the normal population. Below that score, the reverse is the case. By that criterion, our unpaced comparison groups show only 5% in Plant A and 0% in Plant B as being sufficiently high to arouse concern; but among the paced groups the figures are 22% and 16%. Of course, the majority of men on the assembly line are perfectly able to manage their lives quite satisfactorily, and it might even be counterproductive to try to eliminate anxiety altogether. There is, however, a slightly increased risk that paced workers will require assistance.

The major snag in a study of this sort is of course the danger that certain jobs attract certain kinds of people; that paced work does not increase the anxiety of any one person, but only anxious men stay in such a job. There are some details of the results which argue against this possibility; the relationships between anxiety and length of service, or between age, being meticulous, and anxiety, are what they would be expected to be if the job created the anxiety rather than merely selecting anxious people. But the argument cannot be watertight until the same people have been studied before they start such a job, and again after they have done it for some time.

Lastly, the range of technologies studied is small, and it is already clear that it matters what kind of person is being employed. A job with a different kind of pacing, or with workers who are relaxed, and 'working to live, rather than living to work' might give different results. Practical men should therefore be cautious in applying the results; but may find them useful in drawing attention to pacing rather than short cycle time as a possible hazard, and that people may be stressed without feeling any more dissatified.

Sources of stress among machine-paced letter-sorting-machine operators

By J. J. HURRELL, Jr and M. J. SMITH

Applied Psychology and Ergonomics Branch,
Division of Biomedical and Behavioral Science,
National Institute for Occupational Safety and Health,
Cincinnati, Ohio 45226, USA

Sources of job stress among some 3000 machine-paced letter-sorting-machine operators were examined via a questionnaire survey. It was found that one of the major sources of stress for these workers was task dissatisfaction. Perceiving the task of letter sorting to be boring and to lack both challenge and interest was associated with a number of negative psychological and physical states.

1. Introduction

While the majority of studies concerned with machine-paced tasks have been directed towards performance and work-capacity considerations, there is a growing body of evidence linking machine-paced work to undue stress among workers and negative health states (Murphy and Hurrell 1980). However, in spite of this evidence, few studies have undertaken the task of identifying the specific sources of stress (or job stressors) that might be associated with or arise from machine-paced work. Indeed, the question of which particular aspects of machine-paced work serve to foster adverse health consequences has gone unanswered. The current survey study was undertaken in part to provide this type of information in one kind of machine-paced work, letter sorting. The job of Multiple Position Letter Sorting Machine (MPLSM) Operator was chosen for study because, with nearly 30 000 operators, it represents one of the largest work groups in the USA engaged in machine-paced tasks.

2. Methods

2.1. *Subjects*

Questionnaires designed to examine perceived job stressors and to characterize health status were mailed to a sample of 6589 MPLSM operators and to a nearly equal number of non-MPLSM control workers randomly selected from 50 postal facilities. (An earlier paper gives comparisons between the

MPLSM and control workers. This paper deals with relationships between variables of interest within the MPLSM groups.) The 50 survey sites were proportionally sampled on the basis of geographic location and facility size from a pool of all 302 post offices in the continental USA using MPLSM equipment. An initial and follow-up mailing resulted in 3205 completed questionnaire returns from MPLSM operators for a 49% response rate. Of these respondents, 53% were male and 47% were female. The mean age of the sample was 34 years (S.D. = 8·9 years).

2.2. *Measures and model*

A wide variety of job factors, or job stressors, believed to be stress-producing in MPLSM operations were sampled together with an equally large number of strain consequences reflecting emotional, behavioural and health difficulties. Various demographic, personality, and social-support type factors were also sampled for consideration as potential modifiers of the stressor variables and/or the amount of emergent strain produced by any stressor.

In the current analysis, three demographic characteristics (age, sex and education) and measures of five perceived job stressors (equity of pay, organizational satisfaction, task satisfaction, work pressure and cognitive demands) were first used to predict two job-related mood states: (1) content-free job satisfaction; and (2) work-related self-esteem. These mood states were thought to represent intermediate strain responses which are themselves predictive of other strain consequences (Caplan *et al.* 1975). The demographic characteristics and job stressors were then used along with the two mood states to predict three additional kinds of strain: (1) affective states (anxiety and depression); (2) somatic complaints (musculoskeletal and gastrointestinal complaints); and (3) select self-reported physician-diagnosed illnesses (gastristis and hypertension). With the exception of the demographic characteristic and reported illness measures, all indices were factor-analytically derived and composed of Likert-type items. Estimated reliabilities for these indices ranged from 0·60 to 0·97.

2.3. *Statistical analyses*

Stepwise (forward selection) multiple regression was used to analyse the relationship between the predictor variables and each strain consequence. To isolate the independent variables which would yield the optimal prediction equation, a cut-off point was determined by two statistical criteria (see Kerlinger and Pedhazer 1973). First, the overall F ratio had to be statistically significant ($p \langle 0·001$), and second, the parital regression coefficient for the individual independent variable being added had to be statistically significant ($p \langle 0·01$). Below this point, not only is the coefficient insignificant but also the amount of variance contributed by each additional variable (R^2 change) is very small.

3. Results

3.1. *Job-related mood states*

It can be seen from table 1 that six factors were significant in predicting work-related self-esteem (contributing over 34% of the variance) and seven factors were significant in predicting content-free job satisfaction (contributing over 44% of the variance). Task satisfaction, accounting for 31% of the total variance in work-related self-esteem and 39% of the total variance in content-free job satisfaction, was the best predictor of both job-related mood states. Those operators who felt more satisfied with the task of mail sorting (i.e., those who reported the task to be more challenging and interesting) tended to report higher levels of work-related self-esteem (feeling important and successful at work) and more content-free job satisfaction (global. job satisfaction).

Table 1. Predictors of job-related mood states.

Predictor	Work-related self-esteem		
	Multiple R	R^2	R^2 Change
Task satisfaction	0·555	0·308	0·309
Organizational satisfaction	0.555	0·325	0·017
Age	0·557	0·332	0·007
Work pressure	0·586	0·343	0·011
Cognitive demands	0·558	0·346	0·003
Equity of pay	0·590	0·348	0·002

$F = 185·31$; $df = 6, 2079$; $p < 0·0001$

Predictor	Content-free job-satisfaction		
	Multiple R	R^2	R^2 Change
Task satisfaction	0·625	0·390	0·390
Organizational satisfaction	0·650	0·422	0·032
Work pressure	0·662	0·438	0·016
Age	0·664	0·441	0·003
Equity of pay	0·666	0·444	0·003
Education	0·667	0·445	0·001
Sex	0·668	0·446	0·001

$F = 285·02$; $df = 7, 2473$; $p < 0·001$

Organizational satisfaction and work pressure were similarly related to both job-related mood states but uniquely accounted for only an additional 2% and 1% respectively of the remaining variance in work-related self-esteem and 3% and 1% respectively of the remaining variance in content-free job satisfaction. Those operators who reported that they were satisfied with organizational aspects of the postal service (e.g., promotion system and supervision) and those who reported less work pressure (pressure involves having to work too fast, having trouble keeping up and having little task control) also tended to report higher levels of work-related self-esteem and content-free job satisfaction. No other predictor variable accounted for more than 1% of the remaining variance in either job-related mood state.

3.2. *Affective states*

Because work-related self-esteem and content-free job satisfaction were thought to represent intermediate strain responses, they were used along with the three demographic characteristics and five job stressors to predict the remaining strain variables. The results for anxiety and depression can be seen in table 2.

Table 2. Predictors of affective states.

Predictor	Anxiety		
	Multiple R	R^2	R^2 Change
Work-related self-esteem	0·328	0·107	0·107
Content-free job satisfaction	0·357	0·128	0·021
Work pressure	0·372	0·138	0·010
Age	0·381	0·146	0·008
Sex	0·389	0·151	0·005
Equity of pay	0·393	0·154	0·003

$F = 61·37$; $df = 6, 2022$; $p \langle 0·0001$

Predictor	Depression		
	Multiple R	R^2	R^2 Change
Work-related self-esteem	0·301	0·090	0·090
Age	0·315	0·100	0·010
Work pressure	0·336	0·113	0·013
Equity of pay	0·340	0·115	0·002
Sex	0·344	0·119	0·004

$F = 53·93$; $df = 5, 2033$; $p \langle 0·0001$

As table 2 indicates, six factors were significant in predicting anxiety (accounting for over 15% of the variance), and five factors significantly predicted depression (accounting for almost 12% of the variance). Work-related self-esteem, accounting for nearly 11% of the total variance in anxiety and 9% of the total variance in depression, was the best predictor of both affective states. Those operators who reported high work-related self-esteem generally reported less anxiety and depression. Content-free job satisfaction and work pressure respectively accounted for an additional 2% and 1% of the remaining variance in anxiety. Here, more satisfied operators and those reporting less pressure tended to report less anxiety. Age and work pressure each accounted for an additional 1% of the variance in depression. Younger operators and those reporting more pressure tended to report more depression. No other predictor accounted for more than 1% of the remaining variance in either affective state.

Table 3. Predictors of somatic complaints.

Predictor	Gastrointestinal complaints		
	Multiple R	R^2	R^2 Change
Content-free job satisfaction	0·216	0·047	0·047
Sex	0·292	0·085	0·038
Work pressure	0·316	0·100	0·015
Education	0·328	0·108	0·008
Work-related self-esteem	0·332	0·110	0·002

$F = 49 \cdot 80$; df $= 5, 2008$; $p < 0 \cdot 0001$

Predictor	Musculo/skeletal complaints		
	Multiple R	R^2	R^2 Change
Content-free job satisfaction	0·268	0·072	0·072
Sex	0·340	0·115	0·043
Work pressure	0·387	0·150	0·045
Age	0·400	0·160	0·010
Work-related self-esteem	0·410	0·168	0·008
Equity of pay	0·418	0·175	0·007
Education	0·424	0·179	0·004
Cognitive demands	0·425	0·181	0·003
Organizational satisfaction	0·428	0·183	0·002

$F = 47 \cdot 98$; df $= 9, 1928$; $p < 0 \cdot 0001$

3.3. *Somatic complaints*

Table 3 presents the regression results for each of the two somatic complaints scales. As table 3 indicates, five factors were significantly related to the frequency of gastrointestinal complaints (accounting for 11% of the variance) and nine factors were significantly related to the frequency of musculo/skeletal complaints (accounting for over 18% of the variance). Content-free job satisfaction, sex, and work pressure, in combination accounting for 10% of the variance in gastrointestinal complaints and 16% of the variance in musculo/skeletal complaints, were the best predictors of both types of somatic complaint. Less satisfied operators, females and those operators perceiving more pressure tended to report more frequent problems of both varieties. No other predictor accounted for an additional 1% of the remaining variance in either somatic-complaints measure.

3.4. *Reported physican-diagnosed illnesses*

Table 4 indicates that three factors were significant predictors of reported physician-diagnosed high blood pressure (accounting for slightly over 3% of the variance) and that three different factors were significant predictors of self-reported physician-diagnosed gastristis (accounting for about 2% of the variance). Older operators, those reporting greater cognitive demands and

Table 4. Predictors of self-reported physician-diagnosed illnesses.

| Predictor | High blood pressure | | |
	Multiple R	R^2	R^2 Change
Age	0·166	0·028	0·028
Cognitive demands	0·170	0·030	0·002
Work-related self-esteem	0·180	0·032	0·002

$F = 22·56$; df = 3, 2037; $p < 0·0001$

| Predictor | Gastritis | | |
	Multiple R	R^2	R^2 Change
Content-free job satisfaction	0·100	0·010	0·010
Education	0·120	0·015	0·005
Work pressure	0·133	0·018	0·003

$F = 12·29$; df = 3, 2036; $p < 0·0001$

those reporting low work-related self-esteem were most likely to report that they have been told by a physician that they had high blood pressure. Only age, however, accounted for more than 1% of the total variance. Individuals reporting less content-free job satisfaction, less education and greater work pressure were most likely to report a physician diagnosis of gastristis. The best predictor, content-free job satisfaction, however, accounted for only 1% of the variance.

4. Discussion

Of the five perceived job stressors examined, task satisfaction appears to have the greatest impact upon MPLSM operators. Task satisfaction contributes heavily to work-related self-esteem and content-free job satisfaction. These latter job-related psychological states in turn appear to be important determinants of both mental and physical health. Work-related self-esteem, for example, is the best predictor of both anxiety and depression, while content-free job satisfaction is the best predictor of gastristis, as well as gastrointestinal and musculo/skeletal complaints. Thus, dissatisfaction with the actual task of sorting (or more precisely, finding the task to be boring and to lack challenge and interest) seems to have both a direct and indirect strain effect; a direct effect upon occupational self-esteem and global job satisfaction and an indirect effect upon physical health and affective states. Since there is near-universal agreement in the literature concerning the lack of challenge and boring nature of most forms of paced work, task satisfaction may represent a special problem for machine-paced workers.

References

CAPLAN, R. D., COBB, S., FRENCH, J. R., Jr, VAN HARRISON, R., and PINNEAU, S. R., 1975, *Job Demands and Worker Health.* US Department of Health, Education and Welfare, Publication No. (NIOSH) 75–169.
KERLINGER, F. N., and PEDHAZER, E., 1973, *Multiple Regression in Behavioral Research* (New York: Holt, Rinehart & Winston).
MURPHY, L. and HURRELL, J. J., Jr, 1980, Machine pacing and occupational stress. In *New Developments in Occupational Stress,* edited by R. M. Schwartz, DHHS (NIOSH) Publication No. 81–102, Washington D.C.

Stress and health effects in paced and unpaced work

M. J. SMITH, J. J. HURREL, Jr and R. K. MURPHY, Jr.

Applied Psychology and Ergonomics Branch,
Division of Biomedical and Behavioral Science,
National Institute for Occupational Safety and Health,
Centers for Disease Control,
Public Health Service,
Department of Health and Human Services,
Cincinnati, Ohio 45226, USA

The health consequences of machine-paced work have been of interest to occupational health researchers since the inception of these work systems. Hurrell (1981) has presented a comprehensive review of the literature in this area and concludes that machine-paced work appears to have a number of potentially negative health outcomes but that limitations in research methodologies have hampered efforts to draw definitive conclusions. The current study was a large field evaluation of machine-paced letter-sorting workers and other comparison groups of the US Postal Service. The intent was to define, using a questionnaire survey, physical as well as psychological stress problems encountered by these workers.

1. Methods

The questionnaire survey of postal-service employees was carried out at 50 postal worksites distributed uniformly within the USA. Worksites were selected using a stratified random sample by facility size. Within each worksite, employees were randomly selected from employment rosters using a predetermined sampling fraction that would provide the desired number of participants by job category. Six job categories were surveyed; they were multiple-position letter-sorting-machine operators (MPLSM), single-position sorting-machine operators (SPLSM), distribution clerks (DC), electronic technicians (ET), maintenance (MJ) and mechanics (M).

The questionnaire survey instrument was a specially designed form, based in part on previous research study forms (Caplan et al. 1975, Smith et al. 1979) for examining a number of job-specific stress and worker-health areas such as job stressors, health complaints and disease states.

Data from individual questions concerning job satisfaction, job stressors and physical strain were factor-analysed and then combined into scales. These scales were compared across the independent variables of interest, such as job type, using the general linear-model approach to Analysis of Co-variance for unequal cell sizes (SAS Institute, Inc. 1979). In addition, health complaint and disease state were analysed using Mantel Haenszel Analysis to compare the independent variables of interest.

2. Results

The analysis of the demographic make-up of the job types indicated that there were differences between the six job categories in terms of the distribution of workers by age, sex, marital status and race. Based on the extent of differences observed, it was determined that only comparisons between multiple-position letter-sorting-machine operators and distribution clerks could be legitimately made without breaking the data into smaller subsets by age, sex, marital status and race, which was not desirable for this evaluation.

Fifty per cent of the MPLSM operators responded to the survey while 49% of the distribution clerks responded. Table 1 shows the mean response scores for the MPLSM operators and the distribution clerks for the job-stress scales. As can be seen, the MPLSM operators reported greater organizational satisfaction but less job satisfaction, greater cognitive demands, more task dissatisfaction, more work pressure, less coworker support, and less satisfaction with equity of pay.

Each of these job-stress scales was examined in detail to determine the relationship of age, sex and marital status with job. Controlling for age effects, an analysis of co-variance indicated that the interaction of job and sex was significant for work pressure. Male MPLSM operators reported more pressure than female MPLSM operators, while male distribution clerks

Table 1. Mean response scores of MPLSM operators and distribution clerks on job-stress/strain scales.

Stress/Strain Scales	MPLSM Operators	Distribution Clerks
Job satisfaction*	8·9	9·5
Organizational satisfaction*	16·0	15·2
Supervisory support	8·9	9·1
Coworker support*	10·3	10·9
Support of spouse	13·3	13·2
Task dissatisfaction*	20·6	17·9
Cognitive demands*	15·2	14·2
Work pressure*	31·0	26·5
Equity of pay*	13·8	14·1
Fatigue index*	1·9	1·6
Visual disturbances*	1·7	1·3
Muscular problems*	5·6	4·8

* significant at $p < 0.01$ level using Analysis of Co-variance for unequal cell sizes.

reported less pressure than female distribution clerks. No other interactions were significant; however, there were main effects other than job types for some of the stress scales. Females reported higher cognitive demands, greater job satisfaction and greater pay equity than males, while males reported greater task dissatisfaction and more coworker support. In addition, single workers reported less cognitive demands and lower work pressure than married or divorced workers.

The MPLSM operators reported higher levels of physical strain as reflected in three strain scales examining general fatigue, vision disturbances and

Table 2. Percentage of MPLSM operators and distribution clerks reporting select health complaints in the last year showing a significant effect*.

Complaint	MPLSM Operators	Distribution Clerks
Frequent colds	71	65
Skin problems	37	31
Tearful or itching eyes	60	47
Ringing/buzzing in ears	45	36
Severe headaches	41	36
Fainting or dizziness	19	16
Nervousness	40	30
Sweaty/trembly	50	43
Pain down arm	23	19
Fatigue/exhaustion	56	48
Heartburn/acid indigestion	62	55
Gas/gas pains	66	59
Tight feeling in stomach	40	32
Bloated	53	48
Neck pressure	49	38
Trouble with digestion	27	23
Blurred vision	41	31
Dry mouth	44	35
Stomach pains	46	40
Increased urination	40	37
Loss feeling in fingers or wrists	24	16
Neck strain	49	40
Neck pain into shoulder, arm, hand	34	26
Finger cramps when working	38	20
Loss of strength in arms, hands	24	19
Eye strain	69	53
Stiff or sore wrists	39	22
Headaches	80	69
Numbness	28	26

* significant at $p \langle 0 \cdot 01$ level using Mantel Haenszel analysis stratified for age.

muscular problems. In each case, females reported much higher levels than males.

Respondents were asked to report the frequency of occurrence over the past year for 52 separate health complaints. For each health complaint, comparison was made of the percentage of workers in each job within age groupings having experienced a health complaint regardless of the frequency of occurrence (table 2). There were 29 health complaints for which there was a significant difference between the two jobs, and for all of these there were more MPLSM operators reporting a problem. These health complaints dealt mainly with visual, muscular, digestive and psychosomatic problems.

Workers were also asked to report any medical diagnosis of a disease state within the previous five years for 21 separate diseases. Only high blood pressure, for which distribution clerks reported a higher level, demonstrated a significant difference across jobs within age groupings.

3. Discussion

The results of this study indicate that machine-paced mail-sorting-machine operators reported higher levels of job stress than distribution clerks doing mail sorting by hand. Particularly, the stress was related to overall job dissatisfaction and task dissatisfaction, highly repetitive cognitive demands, and high work pressure. In addition, the MPLSM operators reported higher levels of strain as reflected in health complaints. There was little difference in reported disease states except that the distribution clerks reported a higher incidence of high blood pressure than the MPLSM operators. It must be recognized that the response rate to the survey was only 50% and therefore the results might not be representative of the entire workforce of interest.

Essentially, these results confirm the findings of previous researchers that machine-paced work has a negative psychosocial impact on workers and potential health consequences. However, there are a number of issues that must be evaluated before the actual implications of this research can be determined. A very important issue deals with the differences in job tasks between the MPLSM operators and the distribution clerks. While they both sorted mail, the machine operators' job required physical activity such as keying and rigid inspection that differed substantially from the job requirements of the distribution clerks. These job requirements, which imposed more strenuous visual and muscular loads on the MPLSM operators, could be the primary reasons for the higher strain and health complaints observed in MPLSM operators.

To examine whether the strain observed in the MPLSM operators was due to the job requirements related to keying and inspection or those related to pacing and repetitive work, comparisons were made with a select sample of unpaced machine (SPLSM) operators examined. While the demographic make-

Table 3. Percentage of MPLSM operators and distribution clerks reporting a disease state diagnosis by a physician in the last five years.

Disease State	MPLSM Operators	Distribution Clerks
Diabetes	3	5
Cancer	2	1
Hernia/rupture	4	7
Tuberculosis	1	2
Asthma	5	4
High blood pressure*	16	24
Heart disease	3	6
Arthritis	10	8
Epilepsy	1	2
Glaucoma	1	3
Paralysis/tremor/shaking	2	3
Kidney or bladder	14	11
Lung/breathing problem	9	9
Stroke	1	2
Anaemia	9	7
Gall bladder/liver/pancreas	5	5
Thyroid/goitre	4	4
Insomnia	7	7
Gastristis	12	12
Colitis	4	4
Stomach ulcer	6	7

* significant at $p < 0.01$ level using a Mantel Haenszel analysis stratified for age.

up of the groups did not allow for complete comparisons, it was possible to compare male MPLSM operators to male SPLSM operators. Comparisons were made in the type and incidence of health complaints reported, and demonstrated that the male paced MPLSM operators reported significantly higher levels of visual, muscular and other complaints than the SPLSM operators. This analysis was of a limited scope and must be interpreted with caution, but it does support the belief that the types of strain effect observed in the MPLSM operators are as much related to the pacing demands and psychological aspects of the job task as to the physical requirements imposed by keying and rigid inspection tasks.

A second issue that must be examined deals with the lack of differences in reported disease states between the MPLSM operators and the distribution clerks. In fact, for the only disease state that showed a difference (high blood pressure), the distribution clerks reported the higher level, even when the older age of distribution clerks was taken into account. This might indicate that the long-term health risk of machine-paced mail sorting is not substantial, and is in fact relatively minor. However, there are reasons why this conclusion may not be justified. Firstly, the relative rates of reported disease for both the MPLSM operators and the distribution clerks were quite high when compared to other high-risk working populations such as shift workers (Smith *et al.* 1981), especially considering the very young ages of the current study groups (over 2/3 of each group was under 50 years of age). Therefore, the lack of differences between the MPLSM operators and the distribution clerks is not a good assessment of risk, for either group.

Secondly, the MPLSM job has an extremely high turnover rate. There are no statistics available from the 50 plants examined in this regard, but discussions with union representatives and postal-service officials indicate that the MPLSM job turnover rate is very high, and increases substantially for operators beyond 40 years of age. This may be why 2/3 of the MPLSM operators in this study were under age 36. The primary reason given by union officials for the increased turnover rate over 40 years of age was operator inability to keep up with the workrate and work pressure of the MPLSM job. When they can no longer handle the MPLSM job, the workers leave the MPLSM job typically to transfer into a distribution-clerk job. It is very likely that a large percentage of the distribution clerks in the comparison group, particularly in the older age categories, were former MPLSM transfers. However, since this information was not collected in the study, it is not possible to verify this conclusion. In any event, the lack of large differences in disease states between MPLSM operators and distribution clerks may be more a methodological failure of the study to control for work history than the lack of real differences.

Regarding the specific elements of the MPLSM job that proved to be most stressful, the essential problem areas seemed to be most associated with the job-task requirements. A more detailed analysis of these issues can be found in Hurrell and Smith (1981). Modification of these requirements could lead to decreases in worker stress and health complaints. For the main part, it is felt that making the job task self-paced would reduce perceived work pressure and cognitive demands problems. However, the repetitive nature of the work makes it very difficult to deal with the boredom and overall low self-esteem, and high job dissatisfaction observed. Some positive impact may be possible through enhancing the already high level of organizational satisfaction reported by the MPLSM operators and through increased supervisory support. Redesign of the mail-sorting equipment may also be in order to reduce visual and muscular strain. Further study using ergonomic analysis techniques are necessary to determine this need.

References

CAPLAN, R. D., COBBS, S., FRENCH, J. P. R., Jr, VAN HARRISON, R., and PINNEAU, S. R. Jr, 1975, Job Demands and Worker Health. *National Institute for*

Occupational Safety and Health Publication No. 75-160 (Washington, D.C.: Government Printing Office).

HURRELL, J. J., Jr, 1981, *Psychological, physiological and performance consequences of paced work: An integrative review* (Cincinnati, Ohio: National Institute for Occupational Safety and Health).

HURRELL, J. J., Jr, and SMITH, M. J., 1981, *Source of stress among machine paced letter sorting machine operators* (Cincinnati, Ohio: National Institute for Occupational Safety and Health).

SAS Institute, Inc., 1979. *SAS Users Guide* (Raleigh, N.C.: SAS Institute, Inc).

SMITH, M. J., COLLIGAN, M. J., and HURRELL, J. J., Jr, 1979, A review of the psychological stress research carried out by NIOSH 1971-1976. In *New Developments in Occupational Stress,* edited by R. Schwartz (Los Angeles: University of California at Los Angeles).

SMITH, M. J., COLLIGAN, M. J., and TASTO, D. L., 1981, *Health and safety consequences of shift work in the food processing industry* (Cincinnati, Ohio: National Institute for Occupational Safety and Health).

The effect of machine speed on operator performance in a monitoring, supplying and inspection task

By S. H. Rodgers† and M. Mui‡

†Health, Safety and Human Factors Laboratory, Eastman Kodak Company, Rochester, New York 14650, USA

‡Management Services Division, Eastman Kodak Company, Rochester, New York 14650, USA

Production machine operators were responsible for keeping a machine running, including supplying five different roll materials, clearing up to 21 different types of jam, and inspecting units, in packs of 150, for quality problems. Machine speed increases from 580 to 720 units per minute had produced concerns about job-stress levels. Before further increasing the speed to 1000 units/min, an evaluation of the stress factors was done. Machine design restricted the options for increasing the size of supply rolls, so as machine speed increased, loading time also increased and left less time for the main non-paced activity, visual inspection. The operator's perceived stress was related more to concern about being unable to properly inspect the product in the allotted time than to the total-body stress of moving and loading supplies and clearing machine jams.

1. Introduction

One way to improve productivity is to increase production machine speed; but industry is concerned that this should not be at the expense of the machine operator. Each system is unique; increasing machine speed in one situation may be stressful, and in another may be barely noticeable to the operator. If design has not anticipated the needs of the human operator, machine speed increases may not be possible with the same staffing. In this chapter, a case study illustrating the impact of machine speed on operator performance is described. It shows how machine and job design can be altered to permit machine speed to be increased without putting additional stress on the operator.

2. The job

The product was laminated from film, cardboard (end-boards), paper, lead foil, and vinyl. The machine operator's job included the activities shown in figure 1.

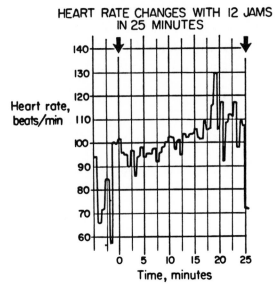

Figure 1. Machine operator's job activities.

1. Keep machine running
   - Keep machines supplied
   - Prepare supply materials
   - Clear jams
   - Unload product in packs of 150
   - Make machine adjustments

2. Do quality inspection on packs of product

3. Other activities
   - Keep records
   - Put product in trays

   - Communicate with colleagues and supervision
   - Anticipate possible maintenance needs

Of these activities, the majority of the shift was devoted to keeping the machine running and doing inspection. Since machine tending was dictated by the needs of the production system, the only discretionary (under the operator's control) activity was inspection. The machine allowed product to accumulate on a conveyor for several minutes (depending on machine speed). The operator inspected product whenever the other machine demands were eased. Product rejected in the machine operator's quality inspection was reinspected at another workstation. The most common defects were incomplete seals and damaged edges. The film in the product required subdued light, so the operator worked in darkroom conditions. Twenty-one different types of machine jam were observed during the study; eight of these occurred more frequently than the others. The machine operators could usually solve the jam problems themselves, but mechanics were available if needed.

3. The question

A new machine was being designed to increase production of this product. The existing machines were running at 580 and 720 units/min; the new machine (and one of the older ones) could be run at 1000 units/min. Departmental supervision had observed that a number of people had difficulty doing the job at 720/minute, although most people could work on the 580/minute machines. The people who were successful on the higher-speed machine liked the job and were challenged by it, but they expressed concern that operating at higher speed would not be possible unless other changes were made. One that they had identified was to change the new machine from a straight-line layout to a horse-shoe configuration, thereby reducing travel time between stations.

We were asked to evaluate the potential impact on the operators of increasing machine speed, focusing particularly on the physical workload and time pressure.

Stopwatch Time, Min.	Elapsed Time Per Activity, Minutes	Activity
20.6	0.2	Start
20.8	1.1	Load Vinyl
21.9	0.9	Inspect
22.8	0.7	Adjust Lead Foil
23.5	1.2	Load Film
24.7	0.7	Inspect
25.4	0.9	Load End Boards
26.3	1.1	Inspect
27.4	0.3	Tray Change
27.7	2.1	Load End Boards
29.8	0.5	Inspect
30.3	1.1	Load Film
31.4	0.3	Inspect
31.7	0.3	Check Film Supply
32.0	0.6	Ticket Work
32.6	1.3	Left Room
33.9	0.6	Die Toggle Jam

Figure 2. Example of activity analysis on machine operator's job.

4. Methods

Four skilled machine operators (male, aged 33 to 44 years) volunteered to participate in the study. Each person's aerobic capacity was predicted from a submaximal treadmill test (Human Factors Section 1982). They were then monitored on the job with a 325 Electrocardiocorder, which made a continuous recording of the ECG throughout the shift (DelMar Engineering 1972). Each person was monitored for two full shifts, one while working on the machine producing 580 units/min, the other on the 720/min machine. During the monitoring a very detailed activity analysis was made of the tasks being performed. This was used to assess the time-stress factor. An example of one portion of this record is given in figure 2.

In addition, the volunteers were encouraged to discuss ways to improve the job and equipment for future designs of higher-speed machines.

5. Results

5.1. *Heart-rate responses*

The heart-rate responses of each operator at the two machine speeds were not significantly different. The range of heart rates was quite narrow, usually from 80 to 105 beats/min (which indicates a moderate workload for this type of work). Lifting the 10 kg lead-foil supply rolls up to the high spindles was difficult for many of the operators, but it was not of sufficient duration or frequency to demonstrate much effect on the heart rate.

One person became frustrated on having 12 machine jams in 25 minutes (see figure 3). This response was attributed more to the circumstances than to a difference in machine speeds.

5.2. *Activity analysis*

The activity analysis provided more insight into the perceived stress on this job than the heart-rate data did. As can be noted from the example in figure 2, machine operators alternated rapidly between activities, preparing and loading supplies, clearing jams, changing trays, inspecting, and doing paperwork. Each activity was performed, on the average, for less than a minute at a time. Stand-by time of a few minutes was rare. The operator was constantly on the move, going from one station to another as needed to keep the machine running smoothly. Each individual had an approach to organizing the work, but all of them demonstrated the short-time-cycle pattern of activities.

As machine speed increases, more time must be devoted to loading and preparing supplies, clearing jams, and inspecting. To keep the pace stress constant, the time available for those activities should increase in proportion to the speed increase. The time to do each direct machine activity (e.g., clearing jams, loading supplies) cannot be reduced substantially, so the activities

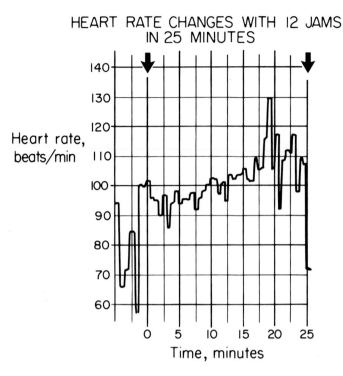

Figure 3. Heart-rate response of an operator when 12 machine jams occurred in 25 minutes.

under the operator's control (e.g., inspection, communications, preparation of supplies) have to be done faster. The activity analysis was used to evaluate the impact of increasing machine speed on each of these activities. Figure 4 summarizes the percentage of time on different activities for the two machine speeds studied, and projects what the impact would be of going to 1000/min.

| | % of Time Over a Shift | | |
Activity	580 units/min	720 units/min	Estimated 1000 units/min
Jam Clearance	9.8	11.3	15.7
Load and Prepare Supplies	24.6	30.6	40.4
Standby	6.7	5.6	4.4
Inspection	21.6	23.4	15.0
Other	37.3	29.1	24.5

Figure 4. Percentage of time on different activities as a function of machine speed.

The category of 'Other' shown in figure 4 refers to communications with colleagues and supervision and machine 'down time' associated with mainten-ance activities or failures. This is generally not time that the operator has control over, so the time stress of the job is better evaluated by the activity pattern during machine operation.

6. Discussion

6.1. *Evaluation of the time stress*

Inspection is the primary task over which the machine operator had discre-tionary control. The product conveyor had to be emptied every 3–5 mins to prevent the machine from shutting down, but did permit the operator to inspect product whenever the demands of other activities were eased. With increasing machine speeds, less time was available for inspection because of the increased loading, jam clearance, and preparation requirements, and because the conveyor filled more quickly. From the activity analysis, we were able to predict the time available for inspection of each pack of 150 units; this is shown in figure 5.

Machine Speed, Units/Minute	Time Available to Inspect Pack, Seconds
580	7.3
720	4.8
1000	1.1*

*Estimated

Figure 5. Impact of machine speed on time to inspect one pack of product.

Quality inspection was difficult at 4·8 seconds per pack, and would be out of the question at 1·1 s/pack. The perceived stress of increasing machine speed from 580 to 720/min, which did not show up on the heart-rate data, was most probably related to the realization by the operators that they could not do a good job under the time constraints of the faster machine speeds. Their approach was two-fold: to look for ways to relieve some of the time pressure on the new production machines (e.g., by design changes); and to reject product that was even slightly questionable, knowing that a second inspection would take place at another location. As a result of the latter decision, they rejected about 10% of the product produced, regardless of machine speed. Reinspection of the rejected product showed that only about 0·5% of it had serious defects.

Another factor that increased time stress at higher machine speeds was the limitation on size of the supply rolls. These rolls were limited in size for three reasons: availability of larger rolls from the supplier (a problem for the vinyl); the spacing between supply spindles on the production machine (not permitting much wider diameter rolls to be used for the vinyl and paper); and the height of the spindles in relation to the weight and size of the rolls (particularly a problem for the lead foil supply rolls). With increased machine speed, there was increased loading required, which could have been reduced if larger supply rolls could have been used.

The length of the machine also increased time stress, since travel times between the loading, inspecting, and most frequent machine jam sites became more critical as machine speed increased.

6.2. *Possible ways to reduce the time stress with increased machine speeds*

Several ways were considered by the machine operators and from the activity analysis to reduce time pressure with increased machine speeds. These can be grouped into machine-design and job-design changes.

Machine-design changes.

(1) Reduce machine length (use horse-shoe configuration) to minimize travel distances.

(2) Lower the supply spindles (especially for lead foil) to about 76 cm.

(3) Increase the space between spindles to permit larger supply rolls to be used; or provide additional spindles to permit the operator to load several supply rolls at once.

(4) Make technological improvements to those locations on the production machine where frequent jams occur.

(5) Increase the amount of conveyor space on the out-coming-product end of the machine so more product can be accumulated between inspection periods.

Job-design changes.

(1) Provide for quality inspection at another station.

(2) Have the machine operator inspect only to identify possible problems with machine adjustment.

Not all of the changes considered could be incorporated into the newer design, because of technological limitations. However, the lead-foil spindle was lowered about 15 cm, the horse-shoe configuration was adopted, and two of the frequent jam sites were improved to reduce the clearance times. In addition, quality inspection was moved to another workstation. Therefore, it is anticipated that the stress on the operator will not be excessive when the new machine is run at capacity.

References

DELMAR ENGINEERING LABORATORIES, 1972, *325 Electrocardiocorder for Continuous ECG Monitoring* (brochure from Avionics Biomedical Division, Irvine, California).

HUMAN FACTORS SECTION, 1982, *Designing for People at Work* (Belmont, Calif: Lifetime Learning) (in the press).

YATES, J. W., KAMON, E., RODGERS, S. H., and CHAMPNEY, P. C., 1980, Static lifting strength and maximal isometric voluntary contractions of back, arm, and shoulder muscles. *Ergonomics,* **23**, 37–47.

Psychoneuroendocrine correlates of unpaced and paced performance

By G. JOHANSSON

Department of Psychology, University of Stockholm,
P.O. Box 6706, S-113 85 Stockholm, Sweden.

Empirical data from a long-term research programme on psychosocial work environment, stress, and health are presented. The results indicate that machine-paced as compared to self-paced work may lead to improved performance, but at the price of a considerable increase in subjective effort and stress and an increase in urinary output of adrenaline and noradrenaline. Since machine-paced work is often combined with other potentially harmful psychosocial factors like repetitiveness, physical constraint, and under-utilization, it is argued that in such cases a restructuring of work is justified for reasons other than the 'pure' effects of machine pacing.

1. Introduction

Recent work reform in Scandinavia is based to a large extent on medical and social science research on man at work. Work Environment Acts and the Act of Codetermination were preceded by thorough analyses of research data concerning effects of work conditions characteristic of industrial production methods (Gardell and Gustavsen 1980, Gardell 1981b). As a result, the new labour law operates within an enlarged work environment concept which includes the content of work as well as traditional health and safety aspects. The implementation of such laws has, in turn, raised new questions to which answers have to be sought in behavioural science.

The data presented here originate from a long-term research programme performed at the Department of Psychology, University of Stockholm, Sweden. The growing recognition of psychosocial risk factors at work provided weighty arguments for combining research strategies from different research areas in a systematic search for potentially harmful factors and high-risk individuals and groups (Frankenhaeuser and Gardell 1976). Theory and methods from social psychology, psychoendocrinology, and experimental psychology were combined. The case of external pacing of work appeared to be an obvious area to be approached experimentally and in the field.

A basic hypothesis of our research has been that quantitative and qualitative overload as well as quantitative and qualitative underload are among the characteristics which generate stress (Zuckerman 1969, Kahn 1973, Frankenhaeuser and Johansson 1981) and—in the long run—may also have a negative impact on physical health. In order to cope with these and other strains it is also necessary for the individual to exert control over his/her immediate environment (e.g., Gardell 1981a, Karasek 1981).

The dependent variables used in our project include self-estimates of wakefulness and mood, and measures of mental performance and of physiological functions. Among physiological parameters the activity of the sympathetic-adrenal medullary system—as reflected in the urinary excretion of the catecholamines adrenaline and noradrenaline—is well suited for our purposes (see reviews by Frankenhaeuser 1975, 1979). Urine samples can easily be obtained in real-life situations without major interference with regular work routines. Adrenaline excretion is a sensitive indicator of behavioural arousal, reflecting both the effort that the individual invests in what he is doing and the intensity of pleasant and unpleasant feelings evoked by the situation. Noradrenaline also reflects behavioural arousal, but usually in a less dramatic way.

Machine-paced work was studied in two laboratory experiments and in a field study. According to the terminology suggested by Salvendy (1981) they were all studies of 'forced feeding'. Both experimental studies used the same complex but repetitive choice-reaction time task, including six visual and two auditory stimuli. All subjects were male students.

2. Empirical evidence

2.1. *Experiment I*

In the first study (Johansson and Lindström 1975), three levels of individual control over work pace were introduced. All subjects performed each of the three tasks for 15 min. This procedure was performed on two separate days, one day at piece-rate remuneration and one day at salary. The remuneration factor will not be analysed in detail here. The following three pacing conditions were used. In a condition named 'no control of work pace' a predetermined stimulus programme was presented to the subjects with the instruction to react as rapidly and as correctly as possible to each stimulus. 'Partial control' was achieved in the same way, but each response by the subject interrupted the prevailing stimulus and elicited a new stimulus cycle. Thus, the rate could be increased but not decreased in comparison with the machine-paced condition. In the 'full-control' condition, finally, the subject generated each stimulus himself by pressing a separate button with his non-dominant hand.

Post-experimental interviews showed that our way of arranging the full-control condition made it more demanding than we had intended because of the extra task of feeding stimuli to the panel.

All the same, as shown by figure 1, individual control was preferred to machine pacing in this task. Heart rate was slightly but significantly lower in the full-control condition than in the other two. Quantitative but not qualitative performance was significantly better in the partial-control condition. Finally, subjective estimates of task duration showed that in the two extreme conditions, no control and full control, time was overestimated. Only when partial individual control of workpace was allowed was the duration of the work period slightly underestimated, indicating that this condition was experienced as less boring than the other two.

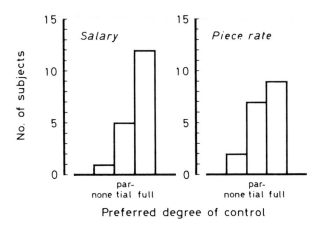

Figure 1. Preferred degree of pacing in a reaction-time task performed in a piece-rate and in a salary session (Johansson and Lindström 1975).

2.2. *Experiment II*

The second experimental study (Johansson 1981) used the same choice-reaction time task but with only two levels of pacing. One group, the 'free-tempo' group, was supplied with an 11-step speed control by which they were allowed to set the rate of signal appearance each 5th min. They were asked to find an optimal work tempo, i.e., one which would result in as many correct responses as possible, but which would also represent a subjectively satisfactory work rhythm.

The 'paced-tempo' group had no speed control. Each paced-tempo subject was randomly paired with a free-tempo subject and given the same sequence of speed as his mate. This arrangement guaranteed that the two groups performed exactly the same task; only the level of pacing differed between them. This group had no other instruction than to react as rapidly and as correctly as possible. Both groups received performance feedback (number of correct responses).

After three hours of work on the choice-reaction time task there was a three-hour period of inactivity for the study of recovery rates of physiological variables. Finally, a second stress condition was introduced in order to reveal possible differences in stress tolerance following paced versus self-paced work. The stressor was an audiovisual conflict test, modelled on Stroop's colour-word test which consists of colour words printed in different colours, the combination of words and colours being incongruent; for example, the word green may be coloured red, etc. The subject's task is to ignore the word and name the colour of the print.

Figure 2 presents the average excretion of catecholamines by the two groups during the work session just described and the control session, which took place on a separate day. The paced-tempo group reached slightly higher levels than the free-tempo group. Although analysis of variance revealed only a significant effect over time for both hormones, the difference between the control condition and the experimental condition (including work) was more obvious for the paced-tempo group than for the free-tempo group.

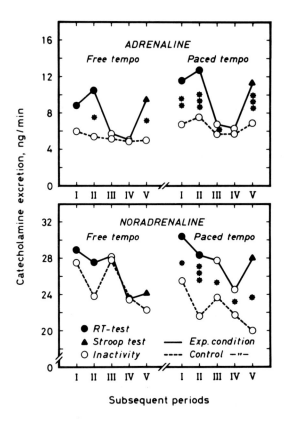

Figure 2. Average excretion of adrenaline and noradrenaline in an experimental work condition and in a control condition for two groups performing a choice-reaction time task. The free-tempo group ($n = 15$) chose their own workpace while the pace of the paced-tempo group ($n = 15$) was predetermined by the pace chosen by matched mates in the free-tempo group (Johansson 1981).

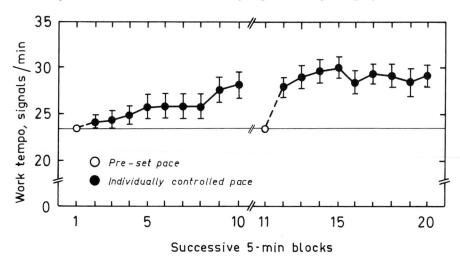

Figure 3. Average workspeed chosen by the free-tempo group in Experiment II (Johansson 1981).

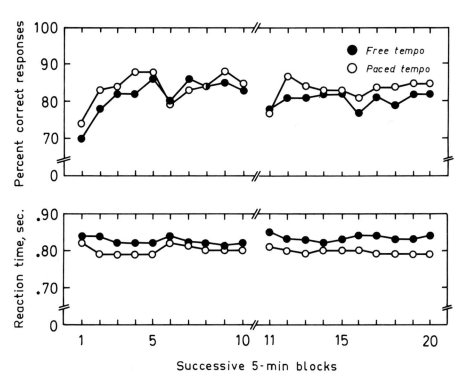

Figure 4. Average performance in a choice-reaction time task performed in free tempo ($n = 15$) and paced tempo ($n = 15$) (Johansson 1981).

A detailed study of the paces chosen by the free-tempo group (figure 3) showed that the initial level set by the experimenter was substituted for successively higher rates until, during Period II, a plateau was reached at about 29 signals/min.

There was a consistent difference in performance between the groups (figure 4). The paced-tempo group performed better in terms of percent correct responses, as well as in terms of reaction time.

Subjective estimates of the stress and effort associated with performing the task are presented in figure 5. Although the work tempo was in fact identical for the two groups, the paced-tempo group reported considerably stronger subjective effort and stress, especially at the end of the work period.

Thus, it can be claimed that the better performance displayed by the paced-tempo group was achieved at the price of increasing effort and stress, and the results illustrate the weakness of performance decrement as a sole indicator of stress.

During the audiovisual conflict test, following the rest period, both groups performed equally well and did not differ in other respects.

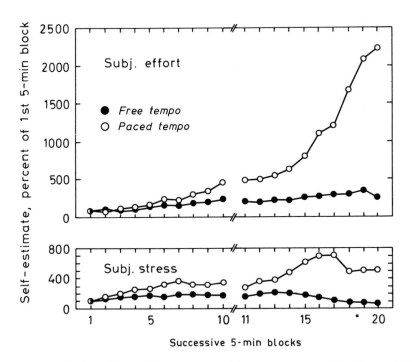

Figure 5. Subjective effort and stress reported by two groups performing a choice-reaction time task in free tempo ($n = 15$) and paced tempo ($n = 15$), respectively (Johansson 1981).

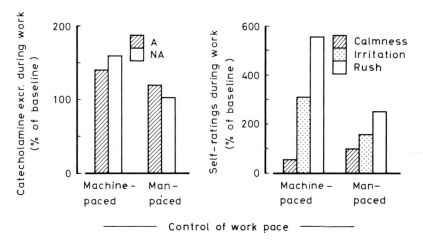

Figure 6. Average excretion of adrenaline (A) and noradrenaline (NA) and self-reported calmness, irritation, and rush, in two groups of sawmill workers performing machine-paced (*n* = 12) and self-paced work (*n* = 12), respectively (Johansson *et al.* 1978).

2.3. *Field study*

Data on machine-paced versus self-paced work were also obtained in a field study of sawmill workers (Johansson *et al.* 1978). When the level of external pacing was related to the output of catecholamines it was found (figure 6) that adrenaline as well as noradrenaline levels over an eight-hour shift was higher for workers in machine-paced than in self-paced jobs. Machine-paced tasks were also associated with higher levels of self-reported irritation and rush.

It should be emphasized that the machine-paced tasks in this case were repetitive, physically constrained and socially isolated. Data from a large-scale ergonomics study of sawmills in Sweden—based on expert ratings, health surveys, and interviews—had shown that groups performing the machine-paced jobs in sawmills reported feelings of extreme monotony and fatigue and that psychosomatic symptoms were exceptionally frequent in this group.

3. Discussion

The studies reviewed showed a tendency towards higher psychoneuroendocrine activity during machine-paced than during self-paced work. They also showed that machine-paced work may lead to an increase in performance, but that this improvement is achieved at the price of considerable increase in subjective effort and stress.

It is important to note, however, that in only one of these studies did the machine-paced task have exactly the same work content as the self-paced task.

And in that case the results should not be generalized beyond the type of repetitive and physically constrained tasks with little opportunity for personal involvement which are available for this type of controlled study.

The problem of separating effects of machine pacing from effects of repetitiveness (short cycle time), constraint, under-utilization of individual resources, etc., has been emphasized in several contributions to this volume and it is an important theoretical and methodological issue. Serious attempts have been made at isolating single factors and their impact on work satisfaction, performance and health (Guest *et al.* 1978, Broadbent 1981). However, from a practical point of view, in developing work-reform strategies, it should be recognized that tasks for which a choice can be made between machine pacing and self-pacing (Mackay *et al.* 1979, Salvendy and Humphreys 1979) usually contain one or more of these other 'confounding' factors. They usually have a short cycle time, they require that the individual stay physically constrained to one place, they allow little autonomy in other respects, and they seldom offer intellectual stimulation.

There is a considerable amount of literature accumulating on the aversive effects of such job characteristics acting separately or in combination. They have been reported to affect job satisfaction, general life satisfaction, stress levels and physical and mental health (e.g., Walker and Guest 1952, Caplan *et al.* 1975, Thorsrud and Emery 1976, Johansson *et al.* 1978, Gardell 1981a, House *et al.* 1979, Mackay *et al.* 1979, Broadbent 1981). It has also been shown how such work conditions may affect the leisure time of the exposed individuals (Meissner 1971, Gardell 1981a, Karasek 1981) so that their participation, especially in cultural and other activities which require active participation and communication with the surrounding world, becomes limited during non-working hours.

It is reasonable to assume that machine pacing plays an amplifying rather than a mitigating role in the development of such unwanted effects. Therefore, whenever machine pacing appears in combination with other potentially harmful psychosocial factors, a restructuring of work should be considered in order to make better use of the intellectual, social and manual resources of the whole human.

References

BROADBENT, D. E., 1981, Chronic effects from the physical nature of work. In *Working Life. A Social Science Contribution to Work Reform*, edited by B. Gardell and B. Johansson (Chichester: John Wiley), pp. 39–51.

CAPLAN, R. D., COBB, S., FRENCH, J. R. P., Jr, VAN HARRISON, R., and PINNEAU, S. R., Jr, 1975, *Job Demands and Worker Health*, Washington: National Institute for Occupational Safety and Health, HEW Publication No. (NIOSH) 75–160.

FRANKENHAEUSER, M., 1975, Sympathetic-adrenomedullary activity, behaviour and the psychosocial environment. In *Research in Psychophysiology*, edited by P. H.

Venables and M. J. Christie (Chichester: John Wiley), pp. 71–94.

FRANKENHAEUSER, M., 1979, Psychoneuroendocrine approaches to the study of emotion as related to stress and coping. In *Nebraska Symposium on Motivation 1978,* edited by H. E. Howe and R. A. Dienstbier (Lincoln: University of Nebraska Press), pp. 123–161.

FRANKENHAEUSER, M., and GARDELL, B., 1976, Underload and overload in working life: outline of a multidisciplinary approach. *Journal of Human Stress,* **2,** 35–46.

FRANKENHAEUSER, M., and JOHANSSON, G., 1981, On the psychophysiological consequences of understimulation and overstimulation. In *Society, Stress, and Disease, Vol. IV: Working Life,* edited by L. Levi (London: Oxford University Press), in press.

GARDELL, B., 1981a, Psychosocial aspects of industrial production methods. In *Society, Stress and Disease. Vol. IV: Working Life,* edited by L. Levi (London: Oxford University Press), in press.

GARDELL, B., 1981b, Strategies for work reform programmes on work organization and work environment. In *Working Life. A Social Science Contribution to Work Reform,* edited by B. Gardell and G. Johansson (Chichester: John Wiley).

GARDELL, B., and GUSTAVSEN, B., 1980, Work environment research and social change: current developments in Scandinavia. *Journal of Occupational Behavior,* **1,** 3–17.

GUEST, D., WILLIAMS, R., and DEWE, P., 1978, *Job design and the psychology of boredom.* London Work Research Unit, Department of Employment.

HOUSE, J. S., McMICHAEL, A. J., WELLS, J. A., KAPLAN, B. H., and LANDERMAN, L. R., 1979, Occupational stress and health among factory workers. *Journal of Health and Social Behavior,* **20,** 139–160.

JOHANSSON, G., 1981, *Individual control in a monotonous task: effects on performance, effort and physiological arousal.* Reports from the Department of Psychology, University of Stockholm (in preparation).

JOHANSSON, G., and LINDSTRÖM, B., 1975, *Paced and unpaced work under salary and piece-rate conditions.* Reports from the Department of Psychology, University of Stockholm, No. 459.

JOHANSSON, G., ARONSSON, G., and LINDSTRÖM, B. O., 1978, Social psychological and neuroendocrine stress reactions in highly mechanised work. *Ergonomics,* **21,** 583–599.

KAHN, R. L., 1973, Conflict, ambiguity and overload: Three elements in job stress. *Occupational Mental Health,* **3,** 2–9.

KARASEK, R., 1981, Job socialization and job strain. The implications of two related psychosocial mechanisms for job design. In *Working Life. A Social Science Contribution to Work Reform,* edited by B. Gardell and G. Johansson (Chichester: John Wiley).

MACKAY, C. J., COX, T., WATTS, C., THIRLAWAY, M., and LAZZARINI, A. J., 1979, Psychophysiological correlates of repetitive work. In *Response to Stress. Occupational Aspects,* edited by C. Mackay and T. Cox (Guildford: IPC Science and Technology Press), pp. 129–141.

MEISSNER, M., 1971, The long arm of the job. A study of work and leisure. *Industrial Relations,* **10,** 238–260.

SALVENDY, G., 1981, Classification and characteristics of paced work. In *Machine Pacing and Occupational Stress,* edited by G. Salvendy and M. J. Smith (London: Taylor & Francis).

SALVENDY, G., and HUMPHREYS, A. P., 1979, Effects of personality, perceptual difficulty, and pacing of a task on productivity, job satisfaction, and physiological stress. *Perceptual and Motor Skills,* **49,** 219–222.

G. Johansson

THORSRUD, E., and EMERY, F. E., 1976, *Democracy at Work* (Leiden: Nijhoff).
WALKER, C. R., and GUEST, R. H., 1952, *Man on the Assembly Line* (Boston: Harvard University Press).
ZUCKERMAN, M., 1969, Theoretical formulations. In *Sensory Deprivation. Fifteen Years of Research*, edited by J. P. Zubek (New York: Appleton Century Crofts), pp. 407–432.

Stress/strain and linespeed in paced work

By L. W. STAMMERJOHN Jr and B. WILKES

Applied Psychology and Ergonomics Branch, Division of Biomedical and Behavioral Science, National Institute for Occupational Safety and Health, Centers for Disease Control, Public Health Service, Department of Health and Human Services, Cincinnati, Ohio, USA

A questionnaire survey was conducted of workers performing a machine-paced inspection task. The survey measured job-related stress/strain on a number of dimensions including job dissatisfaction, workload dissatisfaction, boredom, mood state, and somatic complaints. Responses were analysed for 418 full-time inspectors. Analysis showed that for a number of measures the lowest strain levels in the group were associated with worker-preference line speeds and actual line speeds being approximately equal.

1. Introduction

Over the years, a number of researchers have demonstrated associations between machine-paced work and various forms of physical and psychosocial strain. Adverse psychological effects such as high levels of boredom, anxiety and depression have been indicated for machine-paced workers by Caplan *et al.* (1975), Kornhauser (1965), and Walker and Marriott (1951). Physiological changes associated with paced work have included increased heart rate (Johansson and Lindstrom 1975), elevated catecholamine excretion and somatic complaints (Frankenhaeuser and Gardell 1976), and slow recovery periods after work (Gardell 1980). Salvendy and Pilitsis (1971) have found that relatively young workers show increased energy expenditure under paced work, although older workers appear to experience the opposite effect. Observed worker reactions to paced systems have included temporary work stoppages, strikes and high employee attrition (Belbin and Stammers 1972).

In their 1975 study, Caplan *et al.* adopted a person environment (P–E) fit (French 1973) approach in the study of a number of stress/strain factors. The relationship between the P–E fit and strain may be linear, logarithmic, or curvilinear. Of particular interest to paced work is the U-shaped (curvilinear) relationship.

Clearly, machine-paced work can, at least under some conditions, lead to higher levels of strain than the corresponding self-paced tasks. This study attempts to further determine what relationship exists between measures of strain and the fit of the actual system-controlled workpace (linespeed) and the workpace (linespeed) which the individual would set, given the option.

Specifically, the question of the shape of this relationship, linear or curvilinear, will be addressed.

2. Methodology

The job selected for study was that of machine-paced poultry inspection in the US Department of Agriculture (USDA). The USDA assigns its employees to poultry plants to determine the fitness of their products for human consumption. The majority of the USDA employees in this programme perform post-mortem inspections of poultry carcasses on machine-paced production lines. The inspection process is described in detail by Wilkes and Stammerjohn (1981). The work regimen is strictly paced at rates which are fixed for a given line, and generally range between 15 and 23 carcasses/min per inspector; having little freedom of movement, the inspectors cannot build up or draw on a 'bank' of work by moving up or down the line.

Site visits were conducted at a number of poultry plants to aid in the development of a survey plan and questionnaire for evaluating this activity. During these site visits, some adverse environmental conditions in the plants were observed which included high humidity (rH near 100% in some plants), temperatures which varied widely from plant to plant and from one area of the plant to another, and high noise levels (84–92 dBA). During the site visits the majority of the inspectors were observed standing, although seats were available at most of the workstations. Those inspectors who were observed seated tended to be in awkward postures, e.g., excessive thoracic/lumbar curvature, weight supported by the thighs rather than the buttocks, direction of vision predominantly to one side of the body. From interviews and observations the investigators came to the conclusion that the job involved high visual and musculoskeletal demands related to restricted posture, but low to moderate cognitive demands.

The USDA monitors 240 poultry plants in the continental USA, and from this group a stratified random sample of 121 plants was selected to be included in the study. These plants accounted for approximately 1500 USDA poultry inspection employees out of about 2400 nationally. About 800 of these were full-time line inspectors, with the remainder having some exposure to the inspection task.

A questionnaire survey was developed to examine job conditions and worker health status. The questionnaire included items requesting demographic information, information about specific conditions in the plant, respondent health status, and scales measuring job-related stress and strain.

These latter included three scales tapping job dissatisfaction, workload dissatisfaction and boredom (Caplan *et al.* 1975) and six scales tapping psychological mood (McNair *et al.* 1971). A 53-item symptom checklist (based on the results of the site visits and interviews) was adapted from Tasto *et al.*

(1978). From this checklist six categorical scales were developed, relying on content validity. These were (1) respiratory complaints, (2) musculoskeletal complaints, (3) visual-function complaints, (4) emotional/mood complaints, (5) psychosomatic complaints, and (6) gastrointestinal complaints. A pair of items in the questionnaire asked the respondent to report both the actual linespeed at which the respondent worked and the linespeed which he or she would prefer. From this data, it was possible to compute the individual's PACE-FIT, which is defined as the difference between the actual and preferred number of carcasses inspected per minute by that inspector.

Questionnaires were distributed to all USDA employees in the sample plants who had duties related to poultry inspection. The respondents were instructed to complete the questionnaire at home and return it in a postage-paid envelope which was provided. All responses were anonymous.

The data from all plants in the sample were pooled for analysis. For the purposes of this paper, only the full-time line inspectors were considered. As a first phase, strain scores of these inspectors were tested against PACE-FIT using a Kruskal–Wallis test. As a second phase, the strain scores were fitted to a second-order polynomial using a non-linear least-squares regression with PACE-FIT as the underlying variable. In this test, individuals with PACE-FIT scores outside the range $-3 \leq$ PACE-FIT ≤ 3 were excluded due to the fact that the scale scores could be treated as approximately linear only over a restricted range, which corresponded to the central PACE-FIT region.

3. Results

A total of 638 questionnaires were returned, giving a response rate of 42%. These included 418 full-time inspectors, and 220 others.

Table 1 gives a summary of the mean strain scores for the full-time inspectors by PACE-FIT category. Strain was shown to have a significant relationship to PACE-FIT for 11 of the 15 measures, but this test was inconclusive in determining whether the strain curves were non-linear.

Table 2 gives the results of the non-linear regression of the scores of the full-time line inspectors in the 15 scales versus PACE-FIT. The equation against which the regression was done was a second-order polynomial in PACE-FIT. In seven of the 15 scales, the coefficient of the second-order term was positive ($p(C) > 0$) (< 0.05). In none of the 15 was B, the first-order coefficient, either positive or negative with 95% confidence. The percentages of the sums of the squares accounted for by the regression was quite high in the dissatisfaction scales (82–91%) and for the symptoms scales (88–94%). The percentage of the sums of the squares accounted for was moderate for the psychological mood scales (45–89%), reflecting the multitude of other factors which can effect these measures.

Table 1. Mean reported strain by rating of workpace.

Strain	Too Slow PACE-FIT $\langle\ -0.5$	About right $-0.5 \leq$ PACE-FIT 0.5	Too fast PACE-FIT ≥ 0.5
		PACE	
Job dissatisfaction‡	2·37	2·30	2·62
Workload dissatisfaction‡	2·03	1·87	2·77
Boredom‡	3·01	2·61	3·24
Tension-anxiety‡	7·86	6·56	9·37
Depression-dejection‡	7·31	5·70	8·54
Anger-hostility†	7·00	5·76	8·26
Vigour-activity*	17·05	17·43	17·08
Fatigue-inertia‡	6·87	6·50	9·06
Confusion- bewilderment	5·53	4·65	5·67
Respiratory complaints	1·47	1·63	1·58
Musculoskeletal complaints	1·69	1·49	1·75
Visual-function complaints†	1·41	1·53	1·68
Emotion/mood complaints†	1·47	1·29	1·47
Psychosomatic complaints	1·34	1·27	1·36
Gastrointestinal complaints	1·46	1·45	1·49

* Higher numeric scores on this scale indicate lower levels of strain
† significant at $p\langle 0.05$ by Kruskal–Wallis test
‡ significant at $p\langle 0.01$ by Kruskal–Wallis test

By inspection of the data, it appears that the fit is not as good at the extreme values of PACE-FIT as it is in the central region. This observation is confirmed by the fact that poorer fits were obtained for the regression performed over an extended range.

4. Discussion

The results show that a relationship does exist between PACE-FIT and reported strain. For many of the measures of strain, the effect may be interactive with environmental/ergonomic factors. Of more importance here, however, was the nature of the relationship between PACE-FIT and strain. For

Table 2. Nonlinear regression of measures of strain by PACE-FIT $S = A + B \times$ (PACE-FIT) $+ C \times$ (PACE-FIT)^2.

Strain measure	A	B	C	Percent of sum of squares accounted for
Job dissatisfaction	2·307	0·066	0·44[a]	91
Workload dissatisfaction	2·020	0·098	0·089[b]	82
Boredom	2·709	−0·004	0·139[b]	85
Tension-anxiety	7·051	0·269	0·289[a]	68
Depression-dejection	6·067	−0·116	0·354	45
Anger-hostility	5·976	0·289	0·171	48
Vigour-activity*	17·489	0·253	−0·264	89
Fatigue-inertia	6·858	0·347	0·246	65
Confusion-bewilderment	4·871	−0·256	0·187	67
Respiratory complaints	1·486	0·029	0·018	92
Musculoskeletal complaints	1·625	−0·020	0·041[a]	90
Visual-function complaints	1·519	0·036	0·024	88
Emotional/mood complaints	1·334	−0·007	0·033[b]	90
Psychosomatic complaints	1·303	−0·005	0·011	92
Gastrointestinal complaints	1·427	−0·010	0·021[a]	94

* Higher numeric scores on this scale indicate lower levels of strain
[a] $p(c \leq 0) \leq 0\cdot05$
[b] $p(c \leq 0) \leq 0\cdot01$

seven of the strain measures examined, a pronounced curvature is present in the relationship between PACE-FIT and the reported strain. For the remaining strain measures, no judgement is possible as to the curvature of the strain/PACE FIT relationship.

The curvature of the regression is somewhat dependent upon the PACE-FIT range considered. The curvatures determined by the regression were weaker and inconsistent for $-5 \leq$ PACE-FIT ≤ 5, but were stronger for the range $-2 \leq$ PACE-FIT ≤ 2. In this latter range the psychosomatic complaints scale showed positive curvature (with $p(C)0) \langle 0\cdot05\rangle$. The most probable explanation for this phenomenon is related to the nonlinearity of the strain scales; only over restricted regions of the underlying variable is the assumption of linearity in the outcome variable sufficiently accurate for the nonlinear regression to be meaningful.

The curvilinear (U-shaped) strain/PACE-FIT curves are consistent with the predictions of P–E fit theory as discussed by Caplan *et al.* (1975). This relationship suggests that for machine-paced jobs, a region of minimum strain may exist in the neighborhood of PACE-FIT $= 0$, i.e., if the worker-preferred pace closely approximates the machine-set pace. Causality may not be inferred from this data, and the question remains as to the degree to which the workers' perception of strain may influence the choice of a preferred pace, a phenomenon which might also be related to the extremely high percentage of variance accounted for by the regression. However, it appears to be a reasonable hypothesis that systems which force the worker either faster or slower than this preferred pace may induce strain.

5. Conclusions

The results obtained in this study are consistent with previous findings that a machine-paced work regimen can lead to increased strain. Further analysis shows that the relationship between strain and PACE-FIT follows a curvilinear (U-shaped) model, for at least some strain measures. It should be noted that firm conclusions as to causality cannot be drawn from the regression data, which is similar in its nature to correlational data. It is possible to conclude, however, that the minimum strain levels on many dimensions will be associated with operators working at their preferred pace (PACE-FIT $= 0$). Further research is needed to establish the causality of this relationship and to determine to what extent fitting the pace of machine-paced operations to the worker can ameliorate the strain effects of machine-paced work.

References

BELBIN, R. M., and STAMMERS, D., 1972, Pacing stress, human adaptation and training in car production. *Applied Ergonomics,* **3** (3), 142–146.

CAPLAN, R. D., COBB, S., FRENCH, J. R. P. Jr, VAN HARRISON, R., and PINNEAU, S. R., Jr, 1975, *Job Demands and Worker Health.* US Department of Health Education and Welfare, HEW Publication No. (NIOSH) 75–160.

FRANKENHAEUSER, M., and GARDELL, B., 1976, Underload and overload in working life: Outline of a multidisciplinary approach. *Journal of Human Stress,* September, 34–46.

FRENCH, J. R. P., Jr, 1973, Person role fit. *Occupational Mental Health,* **3** (1), 15–20.

GARDELL, B., 1980, Stress research and its implications in Sweden. *Working Life in Sweden,* **20,** 1–7.

JOHANSSON, J., and LINDSTROM, B., 1975, *Paced and unpaced work in salary and piece-rate conditions,* University of Stockholm, Department of Psychology Report # 359.

KORNHAUSER, A., 1965, *Mental Health of the Industrial Worker: A Detroit Study* (New York: John Wiley).

McNAIR, D. M., LORR, M., and DROPPLEMAN, L. F., 1971, *Profile of mood states.* Educational and Industrial Testing Services, San Diego.

SALVENDY, G., and PILITSIS, J., 1971 Psychophysiological aspects of paced and unpaced performance as influenced by age. *Ergonomics,* **14** (6), 703–711.

TASTO, D. L., COLLIGAN, M. J., SKJEI, E. W., and POLLY, S. J., 1978, *Health consequences of shiftwork.* US Department of Health Education and Welfare, 1978 HEW Publication No. (NIOSH) 78–154.

WALKER, S., and MARRIOTT, R., 1951, A study of some attributes of factory work. *Occupational Psychology* March, 181–191.

WILKES, B., and STAMMERJOHN, L. W., Jr, 1981 *Job demands and worker health in machine-paced poultry inspection.* US Department of Health and Human Services (in press).

Discomfort and motor skills in semi-paced tasks

By I. KUORINKA

Institute of Occupational Health, Helsinki, Finland.

The results obtained from a shop floor and from a laboratory investigation of semi-paced tasks were presented here. The goal of the study was to clarify whether discomfort and skill are related. On the shop floor, the trained and untrained workers rated discomfort differently. Their performance differed as expected. No differences in physiological variables were found. In the laboratory experiment, no clear-cut relation of discomfort to performance times and EMG variables was found. The results give rise to the hypothesis that effort is an important determinant of discomfort. In the laboratory experiment, the I-EMG of a postural muscle showed a daily increasing trend and a variable or a V-shaped course from trial to trial. The cycle times followed a decreasing logarithmic pattern with increasing numbers of pieces.

1. Introduction

Discomfort phenomena are common in semi-paced tasks. They may range from physical and mental fatigue to various collective consequences (Wisner *et al.* 1973). Also, the defined pathological conditions are not uncommon, as has been pointed out by Kuorinka in this volume.

Discomfort may be related to several factors, e.g., postures (Corlett and Bishop 1976) and static work (Kirk and Sadoyama 1973). Postures are usually related to the workplace design, but repetitive tasks themselves rigidify the posture as a sign of the work strain (Teiger *et al.* 1973).

There is clearcut evidence that such skill-related factors as being untrained may cause augmented strain and disorders (e.g., Thompson *et al.* 1951, Wisner *et al.* 1973). Conversely, the strain on workers experienced at work can be low (Hoag *et al.* 1971). Skill relates to accident risk, at least indirectly: the risk of accident is generally higher among those just beginning a job or among those who often change tasks (Hale and Hale 1971).

2. Experiences from packing tasks at a factory

2.1. *Task description*

The tasks in a bread-packing factory are various: feeding the machines; control and adjustment of the flow of the product; and manual packing of the product after the mechanical phase. The feeding tasks are semi-paced; there is a small buffer capacity in the machine. The production line, manipulating the product is continuously running. Not every item of the product, however, requires attention. Thus, this task can also be regarded as semi-paced.

The production line is manned mainly by trained workers, but workers from other departments are occasionally employed to substitute for absentees. Thus, untrained workers occasionally have to fit into a production line, which continues at full speed.

2.2. *Discomfort in trained and untrained workers*

The discomfort rating was gathered from the workers on a five-point scale. The scale points were anchored to attributes which ranged from very 'fatigued' to very 'alert'. The subjects indicated the expression which they felt best suited to their present general feeling. No information of the previous rate was given. Figure 1 gives the ratings from one shift for each worker.

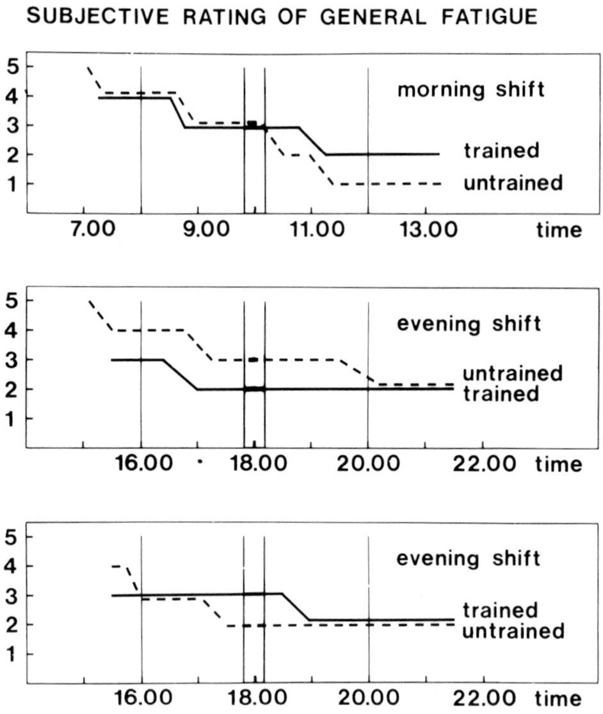

Figure 1. Subjective ratings of three pairs of trained-untrained workers for one shift. The scale is a five-point one. The extremes of the scale were anchored to 'fatigued' and very 'alert'.

The general trend in each pair of rating curves is a declining one, which shows discomfort. In addition to discomfort and fatigue, the curves may show the expectations of each worker about how the rating should go.

To summarize the experiences from the packing factory: there were differences in subjective fatigue between trained and untrained workers. The course of fatigue varied more than the level of fatigue. Taking the breaks in the production into account, the workers may show more of a difference in the expectation than a difference in fatigue or discomfort. The physiological indices showed neither a high level of effort nor any sign of physiological fatigue.

3. Experiences in the laboratory in a simulated assembly task

3.1. *Task description*

The simulation of an assembly task consisted of connecting together pieces of water tubes and yokes by screwing one into the other.

The data on performance were recorded, and the discomfort rating was obtained at the beginning, at the middle, and at the end of the trial, according to the system of Corlett and Bishop (1976). The EMGs of four muscles were recorded as well as working movements on the plane of the working surface.

3.2. *Results of the laboratory experiment*

Performance. The performance of the subjects in the laboratory proceeded as expected. There were long cycle times in the beginning, with a substantial variation. After two trials, the cycle times had reached a level of slow progression.

With the exception of one subject, the best fit of number of workpiece/cycle time curves was logarithmic. The shape of the curves was generally of the same type as found by others, for instance de Jong (1957), Bevis *et al.* (1970), and Bohlen and Barany (1976), although the last two found that exponential and hyperbolic functions best fit the skill acquisition data.

Cycle time curves from the manipulation part of the task differ from those of the tool handling part. Figure 2 shows that the manipulation curves are to a large extent identical for the five subjects. The subjects showed differences in tool handling. The two subjects who learned more slowly and had longer overall cycle times differed considerably from the other subjects. The tool handling included perceptual load (although the task was designed to be predominantly a motor task). As Raouf and Mehra (1974) have shown, informational load increases the cycle times; here, informational load was shown to discriminate subjects with different capacities of skill acquisition.

Physiological strain. The integrated EMG from the right arm and neck muscles showed great interindividual differences. An analysis of the trapezius muscle in the neck showed no clearcut changes when calculated per workpiece. On the contrary, when the I-EMG was calculated for a time unit (1 min), the

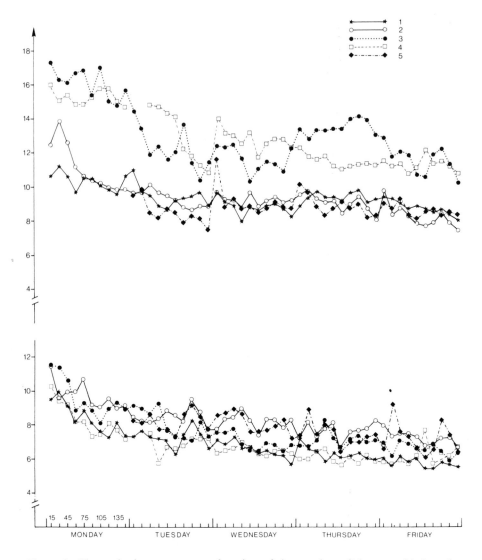

Figure 2. The cycle time curves as a function of the number of the assembled workpieces. There were 150 pieces per day, a total of 750 pieces in five trials (except 600
for one subject). The upper set of the curves describes the cycle times for tool
handling; the lower set shows the cycle time for the manual part of the operation.

curves showed a decreasing trend from one trial to the next, except for one
subject. The daily I-EMG time-unit curves showed an ascending trend
indicative of increased physiological effort or fatigue.

The decreasing trial-to-trial trend was not consistent. In some subjects,
muscle effort increased towards the final trial; thus, least physiological strain

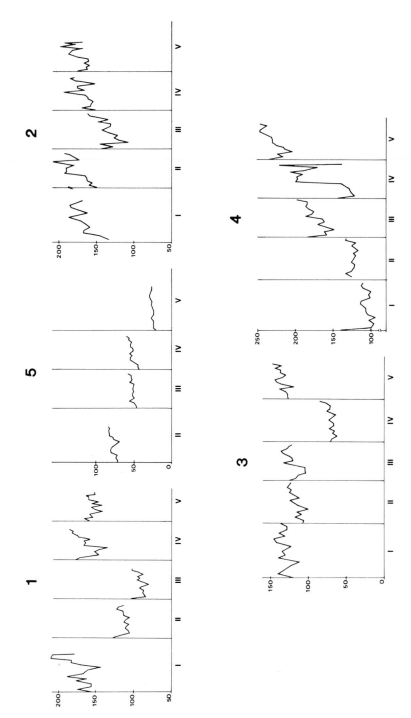

Figure 3. Integrated EMG from m. trapezius (upper part). Individual curves for successive trials are shown.

or effort was exerted in the middle part of the successive trials in three subjects.

Discomfort. In four of the five subjects, the general rating of discomfort showed that the peak rise of discomfort occurred in the third trial. Also in the discomfort rating, individual differences were great, as was anticipated. No difference could be found between the slow learners and the fast learners.

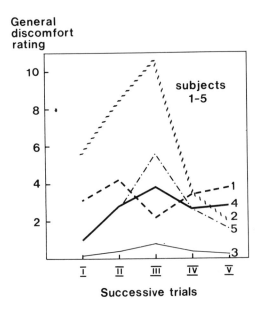

Figure 4. Subjective ratings of general discomfort for the five subjects and consecutive trials. The rating shows the difference between the beginning and the end of each trial.

To summarize the results of our laboratory experiments: we found that cycle time varied in relation to the number of pieces handled. The first cycle times were the longest and had the greatest variability. In most cases, the decline followed a logarithmic function curve.

The physiological strain (I-EMG from one postural muscle) varied or had a V-shaped course from trial to trial. Daily I-EMG curves had mostly an ascending course.

The limited number of subjects does not allow extensive inferences. The results seem to indicate that no direct relationship exists between discomfort and skill. Intervening variables, such as effort, may determine to what extent the lack of skill induces discomfort.

References

BEVIS, F. W., FINNIEAR, C., and TOWHILL, D. R., 1970, Prediction of operator performance during learning of repetitive tasks. *International Journal of Production Research*, **4**, 293–305.

BOHLEN, G. A., and BARANY, J. W., 1976, A learning curve prediction model for operators performing industrial bench assembly operations. *International Journal of Production Research*, **14**:2, 295–303.

CORLETT, E. N., and BISHOP, R. P., 1976, A technique for assessing postural discomfort. *Ergonomics*, **19**:2, 175–182.

DE JONG, J. R., 1957, The effects of increasing skill on cycle time and its consequences for time standards. *Ergonomics*, **1**:1, 51–60.

HALE, A. R., and HALE, M., 1971, A review of the industrial accident research. (London: HMSO). 96p.

HOAG, L. L., HANCOCK, W. M., and CHAFFIN, D. B., 1971, Prediction of physiological strain of workers on the production floor. *International Journal of Production Research*, **9**:4, 457–471.

KIRK, N. S., and SADOYAMA, T., 1973, A relationship between endurance and discomfort in static work. *Ergonomics,* **19**, 2.

RAOUF, A., and MEHRA, M. L., 1974, Experimental investigation related to combined manual and decision tasks. *International Journal of Production Research*, **12**:2, 151–157.

TEIGER, C., LAVILLE, A., and DURAFFOURG, J., 1973, *Taches repetives sous contrainte de temps et charge de travail.* CNAM, Laboratoire de Physiologie du Travail et Ergonomie, Rapport No. 39.

THOMPSON, A. R., PLEWES, L. W., and SHAW, E. G., 1951, Peritendinitis crepitans and simple tenosynovitis: A clinical study of 544 cases in industry. *British Journal of Industrial Medicine*, **8**:150–160.

WISNER, A., LAVILLE, A., TEIGER, C., and DURAFFOURG, J., 1973, *Consequences du travail repetitif sous cadence sur la santé des travailleurs et les accidents.* CNAM, Laboratoire de Physiologie du Travail et Ergonomie, Rapport No. 29.

Psychophysiological studies on stress and machine-paced work

By M. Haider, M. Koller, E. Groll-Knapp, R. Cervinka and M. Kundi

Institute of Environmental Hygiene, University of Vienna,
Kinderspitalgasse 15, A-1095 Vienna, Austria

Occupational stress in machine-paced work is characterized by a combination of monotony and different degrees of mental load. In a series of laboratory and field studies we demonstrated that these combined stressors of boredom and compensatory efforts may lead to different desynchronizations of functions and dysregulations within functional systems. In laboratory studies we pursued the fluctuations and decline of vigilance performance as well as 'cerebral vigilance' by means of EEG frequencies, evoked potentials and slow brain potentials, together with opposed trends of muscle tension (increasing microtremor frequencies) and skin potential responses.

In field studies we found higher pulse rates and longer reaction times in 'secondary tasks' for groups of workers with paced assembly-line work as compared to control groups with self-paced work. This was mainly true at the beginning of working spells, probably caused by the pacing rhythm being especially unphysiological at the onset of work. The dysregulation within the circulatory system was characterized by relatively high pulse rates during machine-paced work, accompanied by relative low systolic blood-pressure values.

1. Introduction

Machine-paced, repetitive work is characterized by a combination of monotony and different degrees of mental load, in a situation with restriction of movement (immobilization), restriction of social interaction and lack of personal control over the entire work process. Repetitiveness, monotony and over-specialization may lead to a 'qualitative understimulation'. Machine pacing and the required continuous, longlasting attention and vigilance may induce overload, overstimulation and hyperactivity, which may be enhanced by physical and chemical environmental stressors like noise, heat and air pollution.

In a series of laboratory and field studies we could demonstrate that these combined stressors may lead to different desynchronizations of functions and dysregulations within functional systems. These changes as well as psychoneuroendocrine stress responses as described by Swedish authors (Frankenhäuser 1977, Johansson et al. 1976) as consequences of machine-paced work in highly mechanized industries may be regarded as threats to health and wellbeing.

2. Methods

In the laboratory experiments, subjects had to perform vigilance tasks, in experimental chambers, lasting 80–90 min. The task required a key-pressing response to dim flashes. The electroencephalogram was recorded from the scalp in bipolar and unipolar modes. Evoked potentials were computer-analysed from the vertex recording site. Pulse rates were calculated from the electrocardiogram. Skin potentials were registered from the palm and the microtremor from the forearm.

In the first field study 32 female workers on assembly lines were compared to 34 female workers who performed piecework. During the study the workers were engaged in a 'secondary' vigilance task. They had to respond as rapidly as possible by depressing a foot pedal whenever small lamps over their work-stations were briefly illuminated. Pulse rates were registered continuously during the whole working spells.

In the second field study 26 female workers with machine-paced work were compared to 28 female workers performing piecework. Pulse rates were measured continuously during the whole working spells (morning and after-noon shifts). Blood-pressure measurements and subjective scalings of work strain were taken before work, every two hours during work and after work. All measurements were taken twice for each female worker on a rotating scheme, once in the morning shift and once in the afternoon shift.

3. Results

In the laboratory studies it could be demonstrated that the different psycho-physiological functions showed different—and in part opposite—tendencies over time. As may be seen in figure 1 (top) the detection efficiency deteriorated over time, and correspondingly the amplitude of computer-analysed evoked responses decreased. Fluctuations of vigilance or attentiveness during the course of the task also were accompanied by corresponding changes in evoked potentials amplitude. Similarly, the EEG-frequencies and the pulse rates were lowered in accordance with the decrease of signal-detection (figure 1, middle). But opposed to these trends, over time the frequency of microtremor and the non-specific skin potential responses increased (figure 1, bottom).

In the first field study, the detection efficiency in the secondary task showed clear differences between paced, assembly-line work and self-paced work. The number of missed signals increased rapidly during paced work, a trend which was interrupted by the longer midday pause. Self-paced work showed no such trend over time. The reaction times were longer for assembly-line work than for self-paced work, especially at the beginning of each work period. This is shown in figure 2.

The pulse rates of the same two groups of female workers are demonstrated

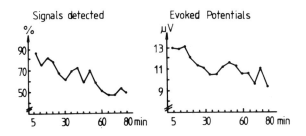

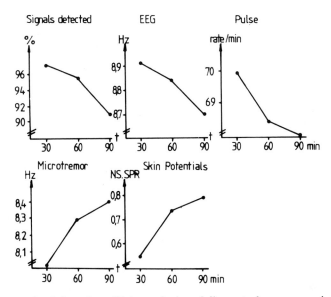

Figure 1. Trends of detection efficiency during vigilance tasks accompanied by downward trends of evoked potentials amplitudes (top) as well as EEG frequencies and pulse rates (middle) but by upward trends in microtremor frequencies and nonspecific skin potential responses (NS.SPR.) (bottom).

in figure 3. The assembly-line workers have a much higher pulse rate than the colleagues with piecework. This is especially true for the morning spell. In the afternoon both groups have a rather high pulse rate.

In the second field study the pulse rates of young female workers (mean age 32 years) were measured during paced and piecework in morning and afternoon shifts. In the morning shifts the pulse rates, starting from an equal baseline value before work, were again markedly increased for paced work compared to piecework. In the afternoon shift both groups have again high pulse rates, but this time even higher values were found in piecework. These findings may be seen in figure 4.

The same groups of workers underwent blood-pressure measurements every

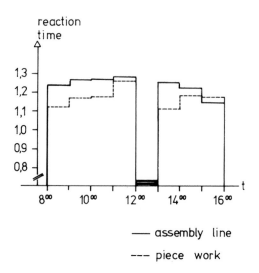

Figure 2. Reaction times in a 'secondary task' (pressing a foot pedal after illumination of a small light) during paced assembly-line work compared to self-paced work.

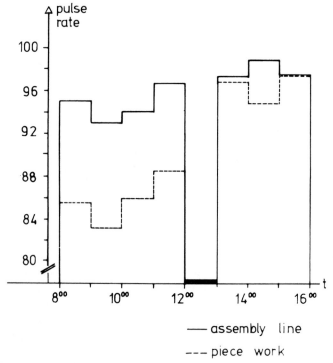

Figure 3. Pulse-rates during paced assembly-line work and self-paced work. The two-hour pulse-rate means are drawn.

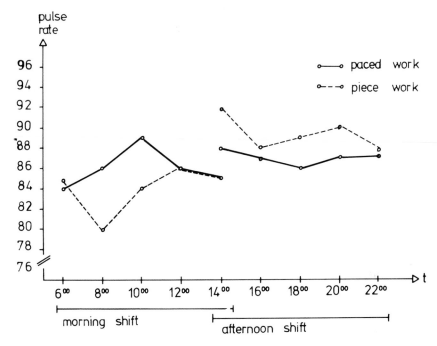

Figure 4. Pulse rates during machine-paced work compared to piecework. Two-hour means are drawn for morning and afternoon shifts.

two hours during their morning and afternoon shifts. The results are demonstrated in figure 5.

The systolic blood-pressure values (top) are rather low for both groups of workers, but they are markedly lower for paced work as compared to piecework. The differences are increased in the afternoon shift. The diastolic blood-pressure values (bottom) remain relatively equal in both groups over morning and afternoon shifts. As a consequence of this, the blood-pressure amplitude is in the afternoon shift much smaller for paced work than for piecework.

4. Discussion

In some of our earlier laboratory studies (Haider *et al.* 1964, Groll 1966), which are re-analysed and summarized in figure 1, we could demonstrate declines of detection efficiency 'cerebral vigilance' (EEG frequencies and evoked potentials) and pulse rates together with opposed (upward) trends of muscle tension and skin potential responses. This may be interpreted as a psychophysiological state between decline of arousal and maintenance of activation, subjectively experienced as a conflict between boredom and compensatory effort. O'Hanlon (1980) has recently shown that this psycho-

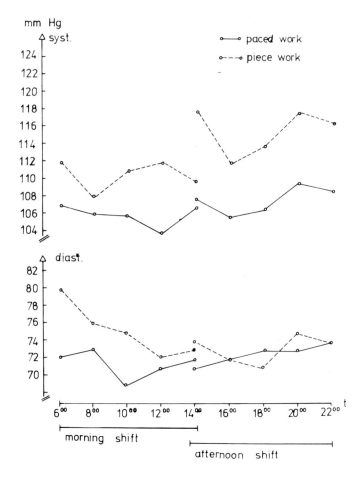

Figure 5. Systolic (top) and diastolic (bottom) blood-pressure values for the groups performing machine-paced work and piecework.

physiological state possesses interrelated and inseparable emotional, motivational, perceptual and cognitive concomitants which may constitute the state of stress.

In our field studies (Haider 1963, Haider and Groll 1966, Cervinka *et al.* 1980) we found higher pulse rates and longer reaction times in 'secondary tasks' for paced assembly-line work as compared to self-paced work. These results fit well with findings of Johansson *et al.* (1976) who found that their machine-paced workers excreted during work significantly more catecholamines than the control group. They also showed a higher frequency of psychosomatic illness and absenteeism than in the control group.

In our studies the differences between paced and self-paced work in pulse rates and response times was most pronounced at the beginning of working spells, especially in the morning. It may be concluded that at this time the effects of the pacing rhythm are most 'unphysiological'. One main course of stress in paced work would then be the fact that there is no possibility of adapting the work to biorhythmic changes.

Concerning circulatory functions, paced-work performed by young women is connected with rather high pulse rates, but rather low systolic blood-pressure values and small blood-pressure amplitude, especially in the afternoon. This may be caused partly through the restriction of movement (immobilization) but may partly also be interpreted as a lack of emotion, motivation and interest in monotonous, repetitive work with 'qualitative understimulation' and without personal control over the work process. All these factors together with the resulting dysregulation of circulatory functions may make paced work a risk factor for disturbances of wellbeing and psychosomatic illness.

References

CERVINKA, R., HAIDER, M., HLOCH, Th., KOLLER, M., and KUNDI, M., 1980. In *Arbeitsbedingte Gesundheitsschäden—Fiktion oder Wirklichkeit* (Stuttgart: Gentner Verlag).

FRANKENHÄEUSER, M., 1977, *Job demands, health and wellbeing.* Reports of the Department of Psychology, University of Stockholm, Nr. 517.

GROLL, E., 1966, Zentralnervöse und periphere Aktivierungsvariable bei Vigilanzleistungen. *Zeitschrift für Experimentelle und Angewandte Psychologie, 13,* 248–264.

HAIDER, M., 1963, Experimentelle Untersuchungen über Aufmerksamkeit und Cerebrale Vigilanz bei einförmigen Tätigkeiten. *Zeitschrift für Experimentelle und Angewandte Psychologie, 10,* 1–18.

HAIDER, M. and GROLL, E., 1966, Belastungsunterschiede bei freier Arbeit und Bandarbeit. *Zeitschrift für Präventivmedizin, 3,* 303–310.

HAIDER, M., SPONG, P., and LINDSLEY, D. B., 1964, Attention, vigilance and cortical evoked-potentials in humans. *Science, 145,* No. 3628, 180–182.

JOHANSSON, G., ARONSSON, G., and LINDSTRÖM, B. O., 1976, *Social, psychological and neuroendocrine stress reactions in highly mechanized work.* Reports of the Department of Psychology, University of Stockholm, Nr. 488.

O'HANLON, J. F., 1980, *Boredom: Practical Consequences and a Theory* (Amsterdam: North-Holland Publishing Company).

Comparison of biochemical and survey results of a four-year study of letter-sorting-machine operators

By R. ARNDT†, J. J. HURRELL‡ and M. J. SMITH‡

†Department of Preventive Medicine,
University of Wisconsin,
Madison, Wisconsin, USA

‡Motivation and Stress Research Section,
National Institute of Occupational
Safety and Health,
Cincinnati, Ohio, USA

Selected results of a four-year study of letter-sorting machine operators are presented. Two groups of workers are compared; the first with three years of experience and the second with two years of experience. Job satisfaction was found to decrease consistently throughout the study while boredom increased. The degree of pacing in the job apparently had the effect of changing workers' attitudes towards pacing, since the amount they would like in their job, if they had a choice, decreased over the years. Moods were found to vary in a variety of ways, with some obviously dependent upon transient events at the workplace rather than years of experience. No consistent results were found for any of the biochemical tests.

1. Introduction

1.1. *Machine pacing and stress*

While machine pacing has been the subject of research for a considerable number of years, few of these studies have been focused upon health consequences. Early studies reported expressions of high job dissatisfaction (Wyatt and Marriott 1951) and 'tension' (Kretch and Crutchfield 1948). Komoike and Horiguchi (1971) reported more frequent health complaints among machine-paced workers and Frankenhaeuser and Gardell (1976) reported psychosomatic problems, including cardiovascular and stress disorders, which were more frequent among paced than unpaced workers. Caplan *et al.* (1975) found that machine-paced assembly-line workers reported the highest levels of job stress of the 23 occupations included in their study. Frankenhaeuser and Gardell (1976) and Johansson *et al.* (1978) found elevated adrenalin secretion in workers involved in highly mechanized paced jobs.

1.2. *Stress among letter-sorting-machine operators*

The majority of mail now passing through the US Postal Service is sorted by letter-sorting-machine (LSM) operators. Letters are mechanically passed in front of the operator who sits at a console. The task, at its simplest level, involves reading two or three of the five digits of the ZIP code and keying them. The console contains 20 keys, ten on an upper keyboard and ten on the lower keyboard. In many cases the task includes memory items or schemes which involve keying something which is missing from the letter or a special sequence of keys for certain types of mail. In many offices, final sorts are performed which place the letters into appropriate carrier routes. The letters are presented at a fixed rate, generally 50 to 60 per minute, although this may be somewhat lower for some special schemes or final sorts. The pacing is rigid, since there is a fixed amount of time during which the letter must be keyed. Operators generally work for 45 min at the console and then work for 15 min at other tasks which include loading the machines and removing the sorted mail.

Reports collected from operators during interviews as well as information provided by union representatives indicated that operators were experiencing a variety of problems which appeared to be job-related. These included psychosomatic complaints such as gastrointestinal problems, headaches and vision problems. Physical complaints included musculoskeletal and nerve disorders in the hands and arms, as well as neck and back pains. In addition, there were reports of increased alcohol consumption, absenteeism and occasional neurotic or psychological disorders. Studies conducted as early as 1967 (Baker) had warned of the potential stressfulness of the job.

Complaints about the job included the 45–15 rotation schedule. These were often intermixed with complaints about certain management policies, as well as hours of work. The majority of operators work night shifts and many have split days off. While these factors appeared to contribute substantially to the complaints of operators, there was enough information to indicate that machine pacing could be playing a major role in the problems being experienced. Finally, there was a concensus that most operators had fewer problems during their earlier years on the job, but that the situation gradually deteriorated, and after about 3–5 years on the job, substantial changes in satisfaction and health occurred.

2. Method

2.1. *Experimental design*

If time on the job is an important factor, there are two ways in which the time course and the levels of stress can be evaluated. The first involves studying workers with varying years of service and the second involves following the same workers over a number of years. The major problem with the first approach is that an accurate picture of the potential stressfulness of the job may be clouded by the lack of information about workers who leave the job. There is also the difficulty of obtaining accurate and.complete information concerning work history. For these reasons, it was decided to conduct a prospective study of workers entering the job.

2.2. *Study site*

A midwestern post office that had recently moved into a new facility and was about to initiate mail sorting by machine was selected for the study. Fifty-four employees of the post office who had no previous experience with the letter-sorting machine, and who had been accepted into the training pro-gramme, volunteered to participate in the initial phase of testing. Over half of these employees never became operators, however. In each subsequent session new employees who had recently become operators were solicited to maintain the sample size at approximately 60 operators.

2.3. *Sample for present study*

Two subgroups were chosen for the present report. The first (G-I) included 31 operators who started working between March and July of 1977 and for whom there was no more than one session of missing data. The second group (G-II) consisted of 21 operators who started working on the LSM during the first six months of 1978. The first group consisted of 25 males and six females. Their average age was 34 and they had an average of $3 \cdot 1$ years of experience as of the last session (August 1980). The second group included 14 males and seven females, with an average age of 31 and $2 \cdot 2$ years of experience as of the last session. The educational backgrounds of the two groups were essentially equal, with 27% high school, 27% some college, and 20% and 10% respectively for the two groups with bachelors degrees. The remaining operators had technical, junior college or graduate training.

2.4. *Procedure*

Each volunteer was assigned a time to come to the testing site during a 3–4 day period. Data were collected once in 1976, 1977, 1978, and 1979 and twice in 1980. Testing included a questionnaire, blood tests, urine tests, and a physical exam.

Questionnaire. Each volunteer completed a questionnaire which included items related to the person, the job environment, job satisfaction, moods, and health.

Blood sampling. Fasting blood samples were taken from subjects for a series of biochemical tests, including uric acid, glucose, cholesterol, triglyce-rides, and creatinine. Tests were performed by a professional laboratory service.

Urine samples. A 24-hour urine sample was collected and analysed for catecholamines.

Physical examinations. A physical examination including blood pressure and heart rate was conducted by a physician.

3. Results

3.1. *Questionnaire results*

An analysis of variance with repeated measures using orthogonal poly-nomials was performed on selected questionnaire data.

Job satisfaction. A scale consisting of five questions was used to evaluate overall job satisfaction. The total range for the scale ran from a low of '0' indicating dissatisfaction to a maximum of '20' for the highest level of satisfaction. The results for the two groups are presented in figure 1. The mean values for the job-satisfaction measure decreased linearly over the years for each of the groups ($p < 0.01$). While Group II consistently scored lower on the job-satisfaction scale, none of the group effects were significant.

Boredom. A scale consisting of four questions was used to evaluate subjective experiences of boredom. The mean values for the two groups are presented in figure 1. The linear increase across years was significant for both groups ($p < 0.01$). Once again there was no significant difference between the groups.

Person–environment fit (pacing). A series of questions were used to separately evaluate workers' perceptions of the extent to which their jobs were paced and the amount of pacing they would like to see in their jobs. The difference between such measures is referred to as the person–environment fit (French 1974). The greater the discrepancy between the two measures, the greater the assumed degree of dissatisfaction would be. The mean values for

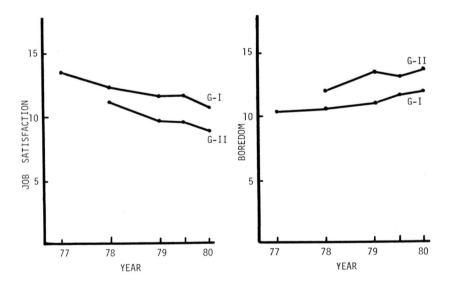

Figure 1. Changes in job satisfaction and boredom.

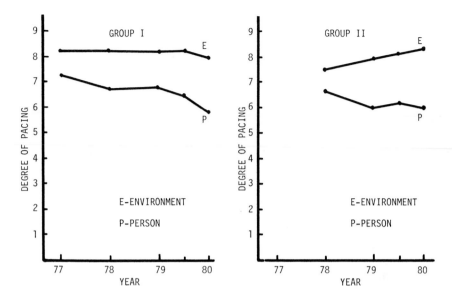

Figure 2. Person–environment fit for Groups I and II.

the perception of their job (E-environment) and what the individual would like in a job (P-person) are shown in figure 2 for Group I and Group II. While the E value for Group I did not change over the years, there was a linear increase in E values for Group II. This is probably due primarily to the fact that many of the Group II subjects were only part-time operators in 1978. Both groups scored significantly lower ($p\langle 0\cdot001$) on the P scale than on the E scale. Furthermore, the P means decreased in a linear fashion over the years for both groups ($p\langle 0\cdot05$).

Moods scales. The profile of moods scores (POMS) for anger-hostility and depression-dejection (McNair, *et al.* 1971) are presented in figure 3. On the anger-hostility scale, Group I increased significantly ($p\langle 0\cdot05$) over the years. Group II, which fluctuated from year to year on the scale, did not exhibit any similar trend.

On the depression-dejection scale there was a significant linear increase in Group II means while Group I showed no change ($p\langle 0\cdot05$).

3.2. *Biochemical results*

Medians for two of the biochemical measures (triglycerides and uric acid) used in the study are presented in figure 4. The yearly pattern of variations shown in figure 4 for triglycerides is typical of the variation seen in a number of other variables, including cholesterol, glucose and creatinine. While some significant changes from year to year were observed, there were no observable trends in the results over the duration of the study. The two groups of workers

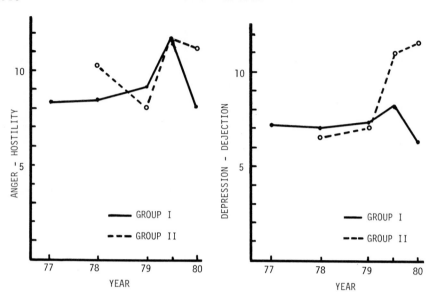

Figure 3. Profile of moods scores for anger-hostility and depression-dejection scales.

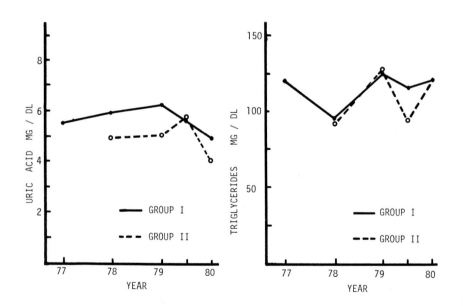

Figure 4. Median values for uric acid and triglycerides.

studied generally tended to vary together. Catecholamine values did not vary significantly from year to year or between groups.

4. Discussion

The results presented above must be considered only as preliminary results, especially where the biochemical results are involved. Nevertheless, some very obvious trends can be seen for the two subgroups examined. Gradually decreasing job satisfaction is apparent in both groups. In the younger and less experienced group this has reached the point of dissatisfaction, in that scores have gone below the neutral point on the scale. A number of factors can be identified which may contribute to this decline in job satisfaction. Two of these are an increase in boredom and a dislike for the degree of pacing involved in the job. The gradual but consistent changes in satisfaction suggest that the length of time on the job is an important consideration. On the other hand, moods appear to fluctuate from year to year and are more likely to be affected by events related to the work situation, such as work performance, management policies, workload, or environmental conditions. A number of situations have been identified, such as changes in shifts and procedures for adjusting chairs, which could account for transitory changes in moods.

At this point, the results of biochemical evaluations do not provide any clear insight concerning the physical effects of changes in moods and attitudes. General effects would appear to indicate that to the extent to which these measures can serve as indicators of stress, they reflect variations in working conditions or situations rather than long-term exposure to stress or job dissatisfaction. Such conclusions must be reserved, however, until more detailed analyses have been completed.

Acknowledgements

I thank Sheri Knutson, Barb Lannigan, Gherry Harding and Kathy Borsecnik for their assistance.

References

BAKER, A. W., 1967, *Exploratory study of selected factors which may influence fatigue and monotony associated with the letter sorting task.* Report to Office of Postmaster General, Southern Methodist University.

CAPLAN, R. D., COBB, S., FRENCH, J. R. P., Jr, VAN HARRISON, R., and PINNEAU, S. R., Jr, 1975, *Job demands and worker health.* DHEW (NIOSH) Publication No. 75-160, NIOSH, Cincinnati.

FRANKENHAEUSER, M., and GARDELL, B., 1976, Underload and overload in working life: Outline of multidisciplinary approach. *Journal of Human Stress,* 2, 35-46.

FRENCH, J. R. P., Jr, 1974, Person role fit. In *Occupation Stress,* edited by A. McLean (Springfield, Illinois: C. C. Thomas), pp. 70-79.

JOHANSSON, G., ARONSSON, G., and LINDSTROM, B. O., 1978, Social psychological and neuroendocrine stress reactions in highly mechanized work. *Ergonomics,* 21(8), 583-599.

KOMOIKE, Y., and HORIGUCHI, S., 1971, Fatigue assessment on key punch operators, typists and others. *Ergonomics*, **14,** 101–109.

KRETCH, T., and CRUTCHFIELD, R. J., 1948, *Theory of Social Psychology* (New York: McGraw-Hill).

McNAIR, D. M., LORR, M., and DROPPLENAN, L. F., 1971, *Profile of Moods States* (San Diego: CA Educational and Industrial Testing Service).

WYATT, S., and MARRIOTT, R., 1951, A study of some attitudes to factory work. *Occupational Psychology*, March, 181–191.

Workers' behavioural and physiological responses and the degree of machine pacing in the sorting of sawmill products

By P. Seppälä and K. Nieminen

Department of Physiology, Institute of Occupational Health,
SF-00290 Helsinki 29, Finland

The error rate, the eye movements, the heart rate and activities not related to work in graders of sawmill products were measured in work conditions which differed according to the degree of machine-pacing and work–rest schedules. Increasing the degree of pacing increased grading errors and heart rate, decreased the rate of blinking and caused a narrowing of the inspected area of the board. Compared to a self-paced work method, a machine-paced work method resulted in more errors, increased the heart rate and reduced activities not related to work.

1. Introduction

The present-day practices of visually grading and sorting sawn timber represent one type of what may be called information-handling tasks (Rohmert and Luczak 1973). The work is usually done in a mechanized setting which involves a compulsory, machine-paced tempo and allows few opportunities for contact with fellow workers (Ager et al. 1975, Seppälä et al. 1979). Tasks performed under such conditions are frequently characterized by mental overload, often combined with monotony (Johansson et al. 1978). More generally expressed, this type of work situation leads to either overstimulation or understimulation. This undesirable state of stimulation has been found to be associated with psychosomatic disturbances and a lack of subjective wellbeing in the workers (Johansson 1975, Gardell 1976).

When the choice between machine-paced and self-paced sorting methods is made, one must determine why the two methods cause different reactions. Is the speed of the conveyor the only important factor, or are there other detrimental effects associated with machine-paced sorting methods? There is some experimental evidence to support the latter interpretation (Poulton 1970, Drury 1973, McFarling and Heimstra 1975). The machine-paced method results in smaller output and more errors in performance, even when the information load is adjusted to equal the information load of a self-paced method. Some studies have found no differences between the inspection performance of workers in machine-paced situations and the inspection performance of workers in self-paced situations (Drury 1973). But, as has been

suggested (Johansson *et al.* 1978), workers may have to pay some psychophysiological cost for working in strictly paced situations.

This paper presents some observations on the issue of pacing, which are
based on two experiments which compared the sorting of sawmill products
both in different degrees of machine pacing and in self-paced versus machine-
paced situations.

2. Methods

2.1. *Subjects*

In the first experiment, 16 graders working in four sorting departments were
studied. The length of experience in sorting work ranged from $1 \cdot 5$ to 28 years
(mean $10 \cdot 2$; S.D. $9 \cdot 8$). Their ages ranged from 26 to 54 years (mean $37 \cdot 9$; S.D.
$9 \cdot 7$). In the second experiment, seven graders working with a modern sorting
machine were studied. The length of experience in sorting work ranged from
one to three years (mean $1 \cdot 7$; S.D. $0 \cdot 9$). Their ages ranged from 26 to 48 (mean
$33 \cdot 7$; S.D. $8 \cdot 0$).

2.2. *Procedures*

In the first experiment, only machine-paced methods were used. Owing to
the different technical construction of the machines, the range of speeds
applied varied somewhat. Three speeds or degrees of pacing were used. The
highest degree of pacing (marked 3) was a conveyor speed which fed 30–46
boards/min; the time available for inspection and decision was $1 \cdot 3$–2 s/board.
The medium degree of pacing (2) was 20–30 boards/min; the time available per
board was 2–3 s. The least degree of pacing was 10–15 boards/min: 4–6
s/board.

Each subject worked a normal workday according to one of three work–rest
schedules: (1) 15 min of grading and 10 min of rest; (2) 30 min of grading,
20 min of butt-end trimming and 10 min of rest; (3) 60 min of grading, 50 min
of butt-end trimming and 10 min of rest. In addition, there were two 20 min
coffee breaks within each schedule.

The graders of the second experiment worked at self-paced tempos and
machine-paced tempos, with two work–rest schedules. The time available for
inspection and decision per board was adjusted to equal the average speed of
the self-paced tempo (3 s/board).

Each subject worked a normal workday according to two work–rest
schedules: (1) 2 h of grading and 20 min of rest (coffee or lunch break); (2) 1/2
h of grading and 10 min of active rest (in addition to the normal coffee and
lunch breaks).

2.3. *Measurements*

In both experiments, grading errors, the number of fixations (EOG) per board, the rate of blinking (EOG), and heart rate were registered. In the first experiment, the width of the inspection area (field of vision in grades) covered by the saccadic movements was also measured from the EOG record. In the second experiment, different subsidiary activities not related to work, such as movements, talking, grasping one's hair, etc. (Kishida 1973), were also observed during the work sessions.

3. Results

3.1. *Grading errors*

Grading errors markedly increased in the first experiment when the highest speeds of the conveyor (30–46 boards/min) were applied ($F(2,18) = 6 \cdot 88$, $p < 0 \cdot 01$) (figure 1). This phenomenon could be clearly noticed when working according to work–rest schedules 1 (15 min period) and 2 (30 min period). With work–rest schedule 3, the greatest percentages of error occurred during the lowest and highest speeds.

In the second experiment, more errors occurred with the machine-paced work tempo than the self-paced work tempo ($F(1,18) = 5 \cdot 58$, $p < 0 \cdot 05$). No significant differences were found between the different work–rest schedules.

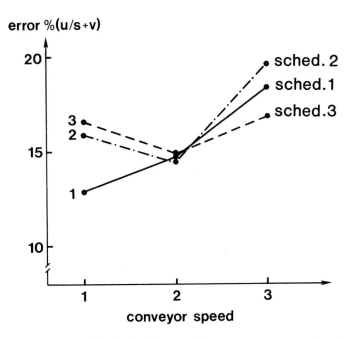

Figure 1. Average error % ($u/s + V$) according to conveyor speed and work–rest schedule.

3.2. *Eye movements*

The field of vision (the area covered by saccadic movements) clearly decreased as the speed changed from low to medium and then remained unchanged or decreased only slightly (figure 2). Also, the number of fixations per board decreased as the conveyor speed increased. The decrease was directly related to the decrease in the time available for inspection (figure 3).

In the second experiment the mean frequency of fixations remained rather stable within the subject group during the entire workday. Thus, it can be concluded that the workers had the same working intensity in self-paced and machine-paced situations, and also in different work–rest schedules.

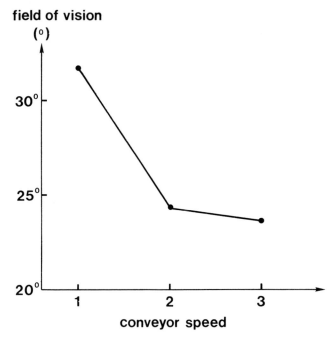

Figure 2. Field of vision (horizontal inspection area covered by saccades) during different conveyor speeds.

The blinking frequency of workers in the first experiment decreased in relation to the increasing conveyor speed or information load ($F(2,24) = 8 \cdot 33$, $p \langle 0 \cdot 01$) (figure 4)

The average blinking frequency in the second experiment was higher in the morning and afternoon periods than in the midday period but the differences between the pacing situations were very small and not statistically significant. One explanation for this is that both the self-paced and machine-paced situations demanded equal or nearly equal concentration.

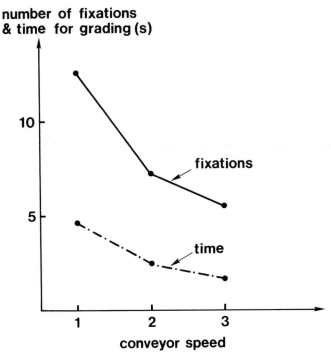

Figure 3. Number of fixations per board and time available for grading during different conveyor speeds.

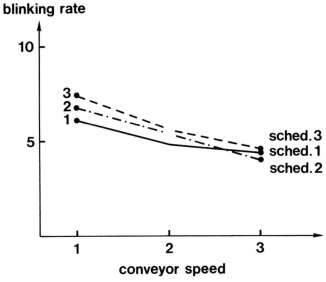

Figure 4. Blinking rate (average number of blinks per 40 s) according to conveyor speed and work–rest schedule.

3.3. *Heart rate*

The effect of the conveyor speed on the heart rate of the subjects of the first experiment is shown in figure 5. The average heart rate increased with increase in conveyor speed $(F(2,28) = 9\cdot23, p\langle0\cdot01)$. The general level was highest in schedule 3.

It is known that, as light physical work continues throughout the day, the heart rate decreases. This phenomenon obviously contaminates the effect of the conveyor speed and must be taken into account when the results are interpreted. Yet, in the first experiment, the average heart rate generally remained throughout the entire workday.

The effect of the time of day and the duration of work sessions was visible in the results of the second experiment (figure 6). The average information load was the same throughout the entire workday and different experimental situations. While the subjects worked according to work–rest schedule 1 (2 h work sessions), the average heart rate generally showed a decreasing trend as the workday progressed $(F(2,30) = 3\cdot47, p\langle0\cdot05)$. This trend was only weakly noticeable in work–rest schedule 2 (30 min work sessions).

The effect of pacing could be seen in the turning points between self-paced and machine-paced situations. When moving from a self-paced situation to a machine-paced situation, the heart rate increased. When the opposite occurred, the heart rate remained unchanged.

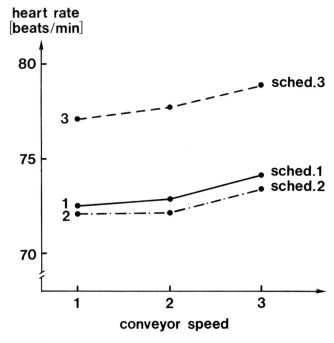

Figure 5. Average heart rate according to conveyor speed and work–rest schedule.

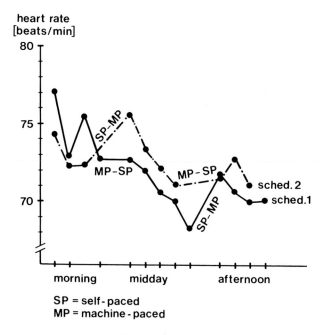

Figure 6. Average heart rate according to time of day and work–rest schedule. The changes between pacing situations are marked in the picture with SP-MP and MP-SP.

3.4 *Subsidiary activities*

More movements and other activities not related to work were found during the self-paced than the machine-paced work situation (figure 7) ($F(1,18) = 3 \cdot 71$, $p \langle 0 \cdot 10$ for movements, $F(1,18) = 5 \cdot 64$, $p \langle 0 \cdot 05$ for other activities). The phenomenon was more clearly noticeable when working according to the work–rest schedule 1, where the periods of grading lasted 2 h without a break, than when working according to the schedule 2, where the periods of grading were interrupted every 30 min by a 10 min active rest pause. (Interaction between pacing and work–rest schedule $F(1,18) = 5 \cdot 205$, $p \langle 0 \cdot 05$ for movements.)

4. Discussion

The results of the grading performance in the first experiment are in agreement with other findings concerning visual search and dynamic visual inspection (Drury 1973). Visual search studies have found that the probability of locating a target is dependent on the amount of time available. The possibilities of speeding up the frequency of fixation without any loss of perceptual

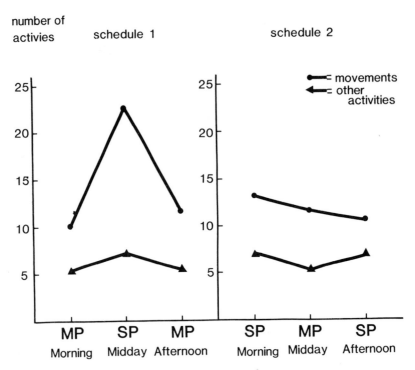

Figure 7. Average number of subsidiary activities according to the work–rest schedule,
      type of pacing and time of day.

efficiency seem to be limited to a certain physiological maximum of about 5
fixations/s (Yarbus 1967). In addition, it is known that dynamic visual acuity
is worse than static visual acuity. Thus, the increased error rate in the present
study can partly be explained by this visual-search effect. Making a decision
after the acquisition of the target, however, is another phase in the process of
grading. Decision-making is affected by the possible changes in criteria which
the worker uses and by changes in sensory discrimination.

The higher percentage of error in the machine-paced work method when
compared to the self-paced one seems to support earlier findings, whereby the
pacing itself results in worse work performance. One possible explanation in
this particular case is that the fixed tolerance time for all boards, independent
of the required search time and the difficulty of the decision, causes the
increased error rate.

The changes in the rate of blinking, heart rate and subsidiary activities can
be interpreted as indicators of increased or decreased effort caused by changes
in information load and demands for concentration and prolonged attention.

The second experiment suggests that, when the demands of the task correspond to the level to which a person has adapted himself, habituation develops,
which leads to a lowered state of arousal.

In order to keep the activation level on the appropriate level for performing the task, the sessions should be interrupted by active rest pauses or other types of work.

References

AGER, B., AMINOFF, S., BANERYD, K., ENGLUND, G., NILSSON, C., SAARMAN, E., and SÖDERQVIST, A., 1975, Arbetsmiljön i sågverk. En tvärvetenskaplig undersönkning. Arbetarskyddsstyrelsen. Undersönkningsrapport AM 101/75.

DRURY, C. G., 1973, The effect of speed of working on industrial inspection accuracy. *Applied Ergonomics*, **4**, 2–7.

GARDELL, B., 1976, *Arbetsinnehall och livskvalitet* (Stockholm: Prisma).

JOHANSSON, G., 1975, *Reaktioner pa överstimulering och understimulering*. Rapporter, no. 6, Psykologiska Institutionen, Stockholms Universitet.

JOHANSSON, G., ARONSSON, G., and LINDSTRÖM, B. O., 1978, Social, psychological and neuroendocrine stress reactions in highly mechanized work. *Ergonomics*, **21**, 583–599.

KISHIDA, K., 1973, Temporal change of subsidiary behavior in monotonous work. *Journal of Human Ergology*, **2**, 75–89.

McFARLING, L. H., and HEIMSTRA, N. W., 1975, Pacing, product complexity and task perception in simulated inspection. *Human Factors*, **17**, 361–367.

POULTON, E. C., 1970, *Environment and Human Efficiency* (Illinois: Charles C. Thomas).

ROHMERT, W., and LUCZAK, H., 1973, Zur ergonomischen Beurteilung informatorischer Arbeit. *Zeitung der angewendeten Physiologie*, **31**, 209–229.

SEPPÄLÄ, P., LOUHEVAARA, V., SIIKARANTA, R., and OJA, P., 1979, *Effect of the production process on mental and physical work demands and strain in the sawmill industry* (Helsinki: Institute of Occupational Health). English summary.

YARBUS, A. L., 1967, *Eye Movements and Vision* (New York: Plenum Press).

Section 7. Impact of computer-paced work on stress

Demands and stress in supervisory tasks: observations in railroad traffic control centres

By C. G. HOYOS

Technische Universität München,
Institut für Psychologie und Erziehungswissenschaft,
Lehrstuhl für Psychologie, FR Germany

The main aim of this investigation was to assess the mental load of job incumbents such as station agents in railway systems and to find out if mental load may impair cognitive functions and initiate human errors.

The PAQ was in our opinion the procedure best suited to identifying in detail demands at the workplace of station agents and to making assumptions about strain of operators. To verify these assumptions, a number of operators were asked to evaluate task elements with respect to strain experienced. This has been done under three aspects: intensity, duration and subjective control. Events at work should be more stressing with increasing intensity and duration and if they are out of control of the operator. Comparisons between ratings by job analysts and evaluations by job incumbents showed a fairly good agreement with respect to stress potential of work conditions.

1. Introduction

The growing interest in occupational stress has to a great extent been stimulated by the rapid changes and developments in communication and control technology and their implementation in industry, traffic, and other areas. This technological change, usually called automation, has generated many new man–machine systems which are more complex and more demanding than older ones. Workplaces in modern man–machine systems are said to induce stress in operators. McGrath (1976) classified the origins from which stress can arise. The most significant origin of stress is to be found in the work task. Tasks which a job incumbent has to perform may be stressing because the task is too difficult in relation to his capabilities, because tasks have to be performed under time pressure or because different aspects of tasks are incompatible with each other. One of the main stressors which has been identified recently is the lack of control workers have over sources of stress, i.e., their lack of ability to predict when a stressor will appear and how intensive its influence will be.

Among modern man–machine systems, those which have been installed in transportation systems, in particular in railway systems, should be given special attention. Control and communication processes in railway systems such as commuter trains in an urban area have to some extent been automated in recent years. Therefore, operating the system no longer entails much manual handling of engines, signals and switches, but instead requires monitoring of automated parts of the system. To do so, operating parameters must be displayed to station agents. This is usually done in control rooms.

To varying extents, direct observation of traffic and handling of controls has been made obsolete by the installation of visual displays and automated controls. This development has made it possible to compare control centres with different levels of automation with respect to their influence in occupational stress.

What makes workplaces in railway systems interesting for a work psychologist? Station agents have to monitor sets of tracks represented on a screen, status of signals, and runs of trains within the section belonging to a certain control centre. The most important characteristic of the monitoring task in question is the dependence of most events to be observed on the schedule of arriving and departing trains. Therefore, the monitoring task which station agents typically have to perform is paced work connected with time pressure. In addition to this main stressor, other conditions such as density of railway traffic in the morning or late afternoon and communications with other operators and offices also contribute to the mental load of operators. All these conditions may create a heavy strain on the operator. Because jobs such as those mentioned before represent demands which can be seen at many workplaces in modern man–machine systems, station agents in railway control centres seemed to be suitable for a study on the aftereffects of stress on human cognitive functions. Here, two aspects of the stress–strain relationship are discussed.

(1) It can be seen very easily from observing job incumbents' activities that operators in control centres are exposed to a number of different stressors. Some of them I have already mentioned. But in stress research, attemps to assess multiple stressors are rare. In many investigations, observations were concerned with one or at best very few stressors at once. There are some other problems in the assessment of stressors. The appropriate description level must be decided on. Would it be better to describe stressors in terms of physical forces such as noise in decibels, or should we rather define stressors in behavioural terms? We decided to describe stressors at work in behavioural terms.

(2) According to Lazarus (1966), McGrath (1976) and other writers, one of the most important aspects of a stress-cycle is human perception of a stressful situation. Attempts to conceptualize occupational stress and its aftereffects in simple input–output relations have not been very convincing. Lazarus (1966)

called this perceptional aspect of stress process 'primary appraisal'.

2. Methodology

To study stressors at work and operators' perception thereof, two instruments were needed: an instrument for assessing stressors and a questionnaire for recording perceived stress.

We considered the possibility of assessing stressors by applying a worker-oriented job-analysis procedure. A procedure which could meet the requirements mentioned above was the 'Position Analysis Questionnaire (PAQ)', which was developed during the last 15 years at the 'Occupational Research Center' of Purdue University by McCormick and his collaborators (McCormick *et al.* 1969, 1972). It consists of about 200 work elements which describe demands at all kinds of workplace in behavioural terms. The PAQ was translated into German and adapted to West German working conditions by Frieling and Hoyos (1978).

The PAQ was originally not designed to be an instrument for assessing stressors. But the following assumption may be warranted: if a job analyst gives a PAQ item a high score on the attached rating scale he will probably have identified a stressor. Accepting this argument, we took the German version of the PAQ, 'Fragebogen zur Arbeitsanalyse (FAA)', for analysing those workplaces in railway control centres which were selected for this study. Psychologists collaborated with officers from West German Railways in analysing positions in question. Each record was discussed carefully with respect to psychological implications as well as to railways service.

If experience of stress depends heavily on perception of environmental conditions, then we should ask people at work how they perceive these conditions. To do so, the following steps were taken: (1) From job demand profiles, 18 work elements were selected which represented varying degrees of demand (14 high-stress ratings, four medium ratings, four low ratings); (2) From original records of job analyses, these aspects of worker activities, for which job analysts gave high ratings, were taken.

Descriptions of these aspects were presented to operators who took part in this study. Here are two examples.

High demand on work element 'Visual displays as sources of job information'. The following statements were given to operator: 'You must observe and interpret continuously the panoramic display in front of you which indicates runs of trains, numbers of trains, position of signals'.

High demand on short-term memory. Following statements were given to operator: 'On your job you must store for a short time numbers of trains, phone numbers, messages about troubles'.

The next step was to formulate questions about operators' experience with these elements of their job. After reviewing a greater number of articles and research reports, Kastner (1978) proposed that experienced strain should be assessed in three dimensions: duration of strain, intensity of strain and controllability of stressors. For each of these dimensions, questions were formulated.

Without doubt, experienced duration and intensity of a stressful condition can be considered to be a valid indicator of operators' strain. So operators were asked: 'How much of your daily work time do you spend observing displayed indicators of train runs?' (for duration) and 'How precisely must you observe the display?' (for intensity). As far as controllability is concerned, it was not so clear what questions should be put to operators. Controllability might be experienced if a person can manipulate a stressing event like noise, or if he can predict a sequence of events (Cohen 1980, Udris 1981). But also, somebody might feel sure that he has things under control if he gets some support from outside in doing this job. Because it seems to be possible to gain control in different ways, a variety of questions about controllability were formulated. A question such as 'How much influence do you have on the sequence of events?' was formulated for controllability in a narrower sense. The question 'How did you become acquainted with such a device?' referred to a more general adaptation to working conditions. With a question such as 'To what extent does this device (rule, procedure) help you in doing your job?' the aspect of support was included.

Writers who introduced the concept of controllability into investigations of stress argued that a lack of control increases the experience of stress, whereas experienced control reduces strain and aftereffects of stress (Frese 1978, Cohen 1980, Udris 1981). However, empirical evidence for this assumption comes from laboratory experiments with noise or from clinical observations. The effect of subjective control on occupational stress is not yet well understood. One main condition may be the extent to which an external condition is fixed.

In most questions which were included in our questionnaire, a high rating on the attached scale indicated high controllability.

A preliminary version of this questionnaire was discussed in detail with representatives of the railway service. After that a final version of the questionnaire was developed, containing 18 descriptions of different tasks of a station agent, with three to five questions on each statement. Members of a research group gave this questionnaire to eight station agents in different control centres. Station agents estimated dimensions of strain in the same way as in the FAA—on a five-point rating scale ranging from low to very high.

As said before, in most questions about controllability, high ratings mean high controllability. Because it has been supposed that subjects experience more stress in the case of low controllability, ratings were transformed into complements of the number six.

3. Results

The main question was: do station agents experience high demands as strain which can be described in terms of duration, intensity and controllability?

Ratings obtained from job analysts as well as from job incumbents were averaged across the different workplaces included in the study. Arithmetic means were ranked in order of magnitude of ratings. Rank correlation coefficients were calculated to answer the question stated above. Table 1 shows the coefficients.

Table 1. Coefficients of rank correlations between experienced strain and estimated demands (*$p < 0.01$; **$p < 0.01$).

	Rho
Duration	0·76**
Intensity	0·50*
Controllability	−0·44
Duration + intensity + controllability	0·42
Duration + intensity	0·79**
Duration × intensity × controllability	0·46

It can be seen that experienced duration of strain agrees strongly with estimated demands. Experienced intensity of strain is in less but still significant agreement with estimated demands. Combined ratings of duration and intensity are in stronger relation to estimated demands. This is in good agreement with what we expected. Results regarding experienced controllability, however, are not in agreement with the expectations formulated above. But because the respective questions represented different aspects of controllability, we considered some subgroups of questions separately. Questions referring to manipulation of stressors and adaptation to stressful conditions were correlated with experts' estimations of demands. We obtained a rank correlation coefficient of Rho $= -0.55$ ($p < 0.05$). This result shows that station agents experienced a feeling of control when work tasks were rated as demanding by experts.

The outcome as to whether an operator felt support from devices, procedures, rules etc., or not—only a few questions referred to this aspect— was in no meaningful relation to levels of demand. Rather, all subjects rated support fairly high.

4. Discussion

In this study one weakness which characterizes a great number of studies could be avoided. External influences, experienced strain, and sometimes aftereffects usually come from the same source, i.e., from job incumbents. We were able to compare levels of demand estimated by job analysts (psychologists and railway officers) with levels of experienced strain estimated by job incumbents. Fairly strong relations between these sets of data allow us to conclude the following.

(1) Levels of demand can be considered as good indicators of task-based stressors.

(2) Operator-experienced stress may depend, besides external conditions, on personal attributes. However, their contribution to the variance of ratings is limited, because station agents are in good agreement about stressing conditions.

(3) Different dimensions of strain are indicated with varying precision by estimated demands, whereby duration is in the best agreement; this might be explained to some degree by the fact that questions about duration have some similarity with the frequency scale of the PAQ.

(4) This and some other studies we have conducted demonstrate the possibility of a modified usage of the PAQ. A careful documentation of activities and conditions which lead to a certain rating of demand allows for the identification of stressing task elements in terms of the job in question. The application of these data in the questionnaire design has been demonstrated in this study.

The results don't support the assumption that high objective and subjective demands are accompanied by the feeling of not having control over things. Rather, reactions of the subjects may indicate successful coping with the many stressors to which they are exposed at their workplace. Strict rules of work conduct, control by office supervisors, even the schedule of train services are obviously not only experienced as a collection of stressful conditions but also as effective instruments for accepting great responsibility.

References

COHEN, S., 1980, Aftereffects of stress on human performance and social behavior: a review of research and theory. *Psychological Bulletin,* **88,** 82–108.

FRESE, M., 1978, Partialisiertes Handeln und Kontrolle: Zwei Themen der industriellen Psychopathologie. In *Industrielle Psychopathologie,* edited by M. Frese, S. Greit and N. Semmer (Bern: Huber), 159–183.

FRIELING, E., and HOYOS, C. G., 1978, *Fragebogen zur Arbeitsanylse (FAA)* (Bern: Huber).

KASTNER, M., 1978, *Entwicklung von Verfahren und Validierung von Indikatoren zur Beanspruchungsmessung bei Kraftfahrern* (Forschungsbericht an die Bundesanstalt für Straßensesen: Köln).

LAZARUS, R. S., 1966, *Psychological Stress and the Coping Process* (New York: McGraw-Hill).

McCormick, E. J., Jeanneret, P. R., and Mecham, R. C., 1969, *The development and background of the Position Analysis Questionnaire (PAQ). Report No. 5* (Lafayette, Ind.: Occupational Research Center, Purdue University).

McCormick, E. J., Jeanneret, P. R., and Mecham, R. C., 1972, A study of job characteristics and job dimensions as based on the Position Analysis Questionnaire (PAQ). *Journal of Applied Psychology,* **56,** 347–368.

McGrath, J. E., 1976, Stress and behavior in organizations. In *Handbook of Industrial and Organizational Psychology,* edited by M. D. Dunne (Chicago: Rand McNally), 1351–1395.

Udris, I., 1981, Streß in arbeitspsychologischer Sicht. In *Streß. Theorien, Untersuchungen, Maßnahmen,* edited by J. Nitsch (Bern: Huber).

Psychosocial factors contributing to job stress of clerical VDT operators

B. G. F. Cohen, M. J. Smith and L. W. Stammerjohn, Jr

Applied Psychology and Ergonomics Branch,
Division of Biomedical and Behavioral Science,
National Institute for Occupational Safety and Health,
Centers for Disease Control,
Public Health Service,
Department of Health and Human Services,
Cincinnati, Ohio 45226, USA.

As a consequence of escalating complaints by video display terminal (VDT) users, the potential health effects associated with VDT use were investigated at five worksites by the National Institute for Occupational Safety and Health (NIOSH). Respondents included: (1) clerical workers using VDTs at either a newspaper or insurance company; (2) clerical workers not using VDTs; and (3) professional newspaper staff using VDTs. This chapter discusses phychosocial factors which could either allay or contribute to job stress of the three groups. The group experiencing the largest amount of job stress and health complaints, clerical VDT workers (CVS), are the primary focus of discussion. CVS' jobs are akin to machine-paced assembly lines in manufacturing plants in the sense that they involve minimal control over tasks or workpace, boring, repetitive tasks, work overload, close monitoring by supervisors, and fear of being downgraded or replaced by the VDT. Controlled studies examining clerical jobs are necessary. NIOSH is planning such field and laboratory studies.

1. Introduction

Interest and concern about machine-paced work has typically focused on industrial assembly lines or shop floors on which workers interact with machines to perform specific, fragmented tasks in the course of producing a product. During the past 30 years, research of this type of work regimen has revealed a number of negative physical and psychological consequences (Caplan et al. 1975, Frankenhaeuser and Gardell 1976, Smith 1981). More recent examinations of these problems are indicated by Ager (1975) and Johansson et al. (1978).

Like mechanized industry, computerization of office work has similar characteristics and outcomes. The use of video display terminals (VDTs) potentially increases the capacity for output and therefore organizational expectations for productivity usually determine a fast workpace. Just as in a machine-paced condition, the clerical worker who must meet excessive production standards has virtually no control over the workpace. Interestingly, 100% of VDT operators in a Swedish insurance company perceived no decrease in the number of routine tasks, the mental strain, or demand for attention since the introduction of VDTs (Johansson and Aronsson 1980). Computerization processes designed to simplify work in order to increase 'thru-put' without concern for human elements turn such offices into clerical assembly lines akin to industrial, mechanized, paced assembly lines.

Over the past few years, the escalation of complaints regarding video display terminals in the USA led to a comprehensive evaluation of the problems by NIOSH which have included: (1) radiation measurements on a sample of VDTs (Murray *et al.* 1981); (2) industrial hygiene sampling of airborne contaminants (Cox 1981); (3) an ergonomic analysis emphasizing illumination/glare problems, workstation design, and operator postural requirements (Stammerjohn *et al.* 1981); (4) an evaluation of job stress and strain reported by operators (Smith *et al.* 1981); and (5) a compilation of health complaints reported by operators (Smith *et al.* 1981). Since many factors of job stress and strain are already examined in some detail in the work of Smith *et al.* (1981) this chapter will focus on the impact that psychosocial factors were found to have on one group of workers engaged in VDT tasks.

2. Method

The investigation was carried out at five workplaces, four newspapers or related operations and one insurance company. The VDT operators at these workplaces were engaged in various jobs including data entry and retrieval, word processing, writing, editing and telephone sales. The control subjects at these workplaces were engaged in data entry and retrieval, word processing, and telephone sales, but did not use VDT equipment in carrying out their work activities. For example, the control subjects doing word-processing tasks were using standard electric typewriters, while those doing data-retrieval tasks were using card-index equipment.

A specially prepared questionnaire was used to gather information about job demands, job stressors, job-stress level, psychological mood, health complaints and working conditions for VDT operators and control groups at each workplace. Various measures of job demands and job stress were contained in the questionnaire. These included scales developed to compare stress in various jobs (Caplan *et al.* 1975), standardized job-stress scales (Insel and Moos 1974), and selected questions on sources of job stress developed specifically for this investigation.

Since the target group for study requested the investigation, traditional sampling strategies and distribution methods were not employed (Smith *et al.* 1981). The respondents to the questionnaire survey were classified into the following three groups on the basis of their work activities: (1) professionals using VDTs; (2) clerical and office workers using VDTs; and (3) clerical workers not using VDTs who served as control subjects. The professionals who used VDTs were mainly reporters, editors, copy editors and printers. The clerical VDT operators included data-entry clerks, data-retrieval clerks, classified-advertizing clerks, circulation and distribution clerks and telephone-inquiry clerks. The non-VDT users were in identical jobs to the clerical VDT operators except that they did not use the VDT in performing their job tasks.

3. Results

The job-stress and psychological-status measures were examined using analysis of variance to compare the mean levels. There were 254 VDT respond-ents to the questionnaire survey for a 50% response rate, and 150 control-group respondents for a 38% response rate. When pertinent data (e.g., amount of VDT usage) was omitted by a respondent, the entire questionnaire was eliminated from analysis. Final groups consisted of 134 professionals using VDTs, 102 clerical VDT users, and 150 clerical controls.

3.1. *General sources of stress*
On eight sources of general life stress and personal problems, significantly more clerical VDT operators reported job stress and health problems than pro-fessionals using VDTs or control subjects, while both clerical VDT operators and professionals using VDTs reported career problems more often than control subjects (table 1).

Table 1. Percentage of VDT operator and control subjects reporting frequent sources of life stress and personal problems.

	Job**	Career*	Finances	Other people	Health*	Time pressures	Unknown	Family
Professional VDT	37	30	26	26	3	41	9	18
Clerical VDT	48	34	31	17	14	37	16	16
Control	29	15	26	14	5	31	10	16

*sign at 0·1 level using a Chi Square test for homogeneity
**sign at 0·05 level

3.2. *Job stress*
All respondents reported elevated stress levels on nine of the ten dimensions of the Work Environment Scale (Insel and Moos 1974) according to estab-lished normative values (table 2). Only the professionals using VDTs did not

Table 2. Mean responses for VDT operators and controls for WES† stress scales

	Involvement**	Peer cohesion*	Staff support	Autonomy*	Task orientation	Work* pressure	Clarity	Control* (by sup.)	Innovation	Physical comfort
Professional VDT (PV)	1·50	2·09	1·85	2·32	1·82	2·25	1·35	1·41	1·24	1·27
Clerical VDT (CV)	1·04	1·33	1·38	1·14	2·02	3·38	1·17	3·04	1·18	1·28
Control (C)	1·25	1·88	1·56	1·75	1·81	2·43	1·55	2·65	0·94	1·13
Duncan Range Results	PV>CV	PV&C>CV	PV>CV	PV>C>CV		CV>PV&C		PV>C>CV		
WES Norms†	2·80	2·73	2·94	2·69	2·51	1·77	2·33	2·32	2·40	2·04

*sign at 0·01 level using an analysis of variance
**sign at 0·05 level
†Work Environment Scale (Insel and Moos 1974) Form S.

show stress levels higher than normative value on the dimension indicating control by one's supervisor. These observations concerning the higher than normative levels of stress for all three groups were not based on statistical tests but solely on examination of the mean values and judgements about their differences.

Using analysis of variance methods and subsequent Duncan range tests for comparing the three groups, the clerical VDT operators (CVS) reported significantly less peer cohesion and job autonomy with more work pressure and greater control by their supervisor than professionals using VDTs (PVS) or control subjects. The CVS also reported less involvement and staff support than the PVS. The professionals who used VDTs reported greater autonomy and less control by their supervisor than the control subjects. For all of the significant stress factors, there was a similar pattern of response in that the CVS reported the highest levels of stress, followed by the control subjects, with the PVS showing the lowest levels of stress.

The response means for the VDT operators and control subjects for nine job demands dimensions developed by The Institute for Social Research (ISR) (Quinn and Shepard 1974, Caplan *et al.* 1975) indicate that the clerical VDT operators reported higher workload, more boredom, greater workload dissatisfaction, greater job-future ambiguity and less self-esteem than either the professionals using VDTs or the control subjects. The CVS also reported more role ambiguity and more boredom than the professionals. The PVS reported the least boredom, workload dissatisfaction, role ambiguity, and job-future ambiguity of all the groups. The same general pattern of stress response observed for the Work Environment Scale was also seen for the ISR stress dimensions, with the clerical VDT operators showing the highest stress levels followed by the control subjects and the professional VDT operators.

Specific job stressors for which the CVS reported more problems than both the PVS and the control subjects fall into four general categories, including problems with workload, workpace, boring job tasks, and lack of career development.

4. Discussion

The results of the study indicate that all worker groups evaluated, including the control subjects, reported high levels of psychological job stress compared to other similar occupational groups (Insel and Moos 1974, Caplan *et al.* 1975). Organizational factors such as strained management–employee relations may have heightened stress levels in these worker groups. Although this specific factor was not measured in the survey, strained relations produced by difficult labour negotiations could have accounted for the increased stress. The control subjects in particular were working under circumstances that could have increased their overall job-stress level. At one site many of the

Table 3. Mean Responses for VDT operators and controls for job demands† stress scales.

	Workload* dissatisfaction	Boredom*	Role** ambiguity	Quantitative workload-Q*	Quantitative workload-E	Lack of self-esteem*	Role conflict	Workload variance	Job future ambiguity*
Professional VDT (PV)	2·21	2·09	1·55	3·55	3·38	9·96	1·82	2·74	3·04
Clerical VDT (CV)	3·17	3·36	1·79	4·04	3·55	12·41	1·71	3·02	3·50
Control (C)	2·43	2·63	1·73	3·61	3·60	9·73	1·92	2·87	3·10
Duncan Range Results	CV>PV&C	CV>C>PV	CV>PV	CV>PV&C		CV>PV&C			CV>PV&C
Job demands and worker health study Median scores†	2·13	1·83	2·06	—	3·51	—	1·75	2·81	2·70

*sign at 0·01 level using an analysis of variance
**sign at 0·05 level
†Scales taken from Job Demands and Worker Health (Caplan *et al.* 1975)

control subjects were aware that they might lose their job by the end of the year due to a business slowdown, and that those who would be retained would become VDT operators. Also, at the other four sites, the control subjects knew they would be converting to VDTs within months. Such factors most likely contributed to an elevation in stress for the control subjects.

The fact that there was only a slight positive correlation between the amount of time of VDT use and the total number of health complaints signifies the need for further examination of the job content and associated psychosocial factors. The pattern emerging from the results clearly indicates that the clerical VDT operators (CV) report the highest stress level, the professional operators (PV) report the least amount and the clerical workers who do not use VDTs fall in between. This suggests VDT use is not the only contributor to job-stress elevation; job content must also be a contributor.

Examining job content, it was found that the CVS held jobs involving rigid work procedures with high production standards, constant pressure for performance, very little operator control over job tasks, and little identification with, and satisfaction from, the end-products of their work activity. In contrast to the CVS, the PVS held jobs that allowed flexibility, control over job tasks, utilization of their education and a great deal of satisfaction and pride in their end-products. Although both jobs had tight deadline requirements, only the PVS had a great deal of control over how these would be met. The CVS viewed the VDT as part of a new technology that took meaning out of their work, increased their workload, and increased their anxiety of being replaced by the very machines they had learned to master, whereas the PVS perceived the VDT as a tool that could be used for enhancing their end-product. This investigation supports the findings of Johansson and Aronsson (1980) that the jobs that profit from automation are those that incorporate latitude with which to regulate one's work arrangement. According to Johansson (1981), Frankenhaeuser and Rissler (1970) and Frankenhaeuser (1979) a significant mediator of stress is the amount of control an individual has over his/her work. Gardell (1979) and Karasek (1979) also indicated that impact of stressors like overload and underload could be assuaged by the extent to which a worker has control over the job.

Lack of task control was previously found to be a job stressor in the following study. Johansson *et al.* (1978) revealed that sawmill workers who had some control over their workpace had lower catecholamine levels, and those who lacked workpace control had higher levels and experienced feelings of rush and irritability.

Again, we see a parallel from assembly-line plant workers to assembly-line-type office workers. The repetitive, paced work demanding concentration that results in monotony and mental overload which could elicit 'the recurrent activation of biochemical adjustment mechanisms...' described by Johansson *et al.* (1978) may also effect the VDT clerical workers' health and wellbeing.

Another striking similarity to traditional assembly lines is the constraint of movement, the limited mobility and fixed posture required of CVS performing a heavy workload of repetitive tasks. Johansson *et al.* (1978) indicated that noradrenaline output, which is influenced by body posture, was low for workers who could move about and high for those remaining in the same place. Johansson and Aronnson's (1980) results of self-reports showed that calmer, more positive feelings are also correlated with mobility. Clerical VDT workers labour under a technology system which disregards the human element in the same way that the assembly line in manufacturing processes led to dehumanizing those work activities. Fear that further office automation will lead to job downgrading or termination has increased job ambiguity and security. This study has shown that the CVS have much less control over boring and repetitive tasks, experience greater dissatisfaction with an increased workload at too fast a pace, and have their work more closely monitored than traditional clerical workers and also more than machine-paced workers in other NIOSH sponsored studies (Caplan *et al.* 1975, Quinn *et al.* 1974).

Working under these kinds of pressure is not conducive to peer cohesion, staff support, or to enhancing self-esteem—all important buffers to negative consequences of stress. Consequently, the VDT clericals scored significantly higher on almost every item in the psychosocial stress scales than professionals using VDTs or than fellow clerical workers not yet using VDTs. Based on this investigation, it is concluded that for the health and wellbeing of the worker, it is essential to examine effects of job content, task requirements, and workload as well as of the workstation design and of the surrounding environment.

References

AGER, B., 1975, A review of the ergonomic problems in sawmills and woodworking industries. In *Ergonomics in Sawmills and Woodworking Industries*, edited by B. Ager (Stockholm: National Board of Occupational Safety and Health), pp. 1–10.

CAPLAN, R., COBB, S., FRENCH, J. R. P., VAN HARRISON, R., and PINNEAU, R., 1975, *Job demands and worker health,* National Institute for Occupational Safety and Health Publication No. 75-160, Washington D.C.

COX, C., 1981, *Potential Health Hazards in VDTs: Industrial Hygiene* (in the press).

FRANKENHAEUSER, M., and GARDELL, B., 1976, Underload and overload in working life. A multidisciplinary approach. *Journal of Human Stress*, **2**, 35–46.

FRANKENHAEUSER, M., 1979, Psychoneuroendocrine approaches to the study of emotion as related to stress and coping. In *Nebraska Symposium on Motivation 1978*, edited by H. E. Howe and R. A. Dienstabier (Lincoln: University of Nebraska Press), pp. 123–161.

FRANKENHAEUSER, M., and RISSLER, A., 1970, Effects of punishment on catecholamine release and efficiency of performance. *Psychopharmacologia*, **17**, 378–390.

GARDELL, B., 1979, *Tjanstemannens arbetsmiljoer (Work environment of white-collar workers)*, Preliminary report, the research group for social psychology of work, Department of Psychology, University of Stockholm, Report No. 24.

INSEL, P., and MOOS, R., 1974, *Work Environment Scale—Form S* (Palo Alto, California: Consulting Psychologist Press).

JOHANSSON, G., 1981, *Individual control in monotonous task: Effects on performance, effort, and physiological arousal*. Reports from the Department of Psychology, University of Stockholm (in preparation).

JOHANSSON, G., ARONSSON, G. and LINDSTRÖM, B., 1978, Social psychological and neuroendocrine stress reactions in highly mechanized work. *Ergonomics*, **21**, 583–599.

JOHANSSON, G., and ARONSSON, G., 1980, *Stress reactions in computerized administrative work*. Stockholm.

KARASEK, R. A., 1979, Job demands, job decision latitude, and mental strain: Implications for job redesign. *Administrative Science Quarterly*, **24**, 285–308.

MURRAY, W. E., MOSS, C. E., PARR, W. H., 1981, Potential health hazard of VDTs: radiation. *Human Factors* (in the press).

QUINN, R. P., and SHEPARD, L. J., 1974, *The 1972–1973 Quality of Employment Survey* (Ann Arbor, Michigan: University of Michigan).

SMITH, M. J., COHEN, B. G. F., STAMMERJOHN, L. W., and HAPP, A., 1981, An investigation of health complaints and job stress in video display operations. *Human Factors* (in the press).

STAMMERJOHN, L. W., Jr, SMITH, M. J., and COHEN, B. G. F., 1981, Evaluation of workstation factors in VDT operations (in the press).

CRT display terminal task pacing and its effect on performance

By A. Raouf, S. Hatami, and K. Chaudhary

Department of Industrial Engineering, University of Windsor,
Windsor, Ontario, Canada N9B 3P4

The effect of five baud rates on time taken and errors committed in editing a simple text using four types of alphanumeric CRT display terminals was studied. Ten subjects participated in the experiment. A $5 \times 4 \times 10$ factorial design was used. Baud rate was found to be a highly significant factor. An optimal baud rate of 1200 bits per second was found. At this baud rate, time taken to edit the text and the number of errors committed were minimum.

1. Introduction

It is generally believed that office workers comprise nearly 22% of the labour force, but account for 40–50% of the manufacturing costs. In the past, industrial engineers and others have tried to improve the productivity of manufacturing operations. By one estimate, between 1968–78 the USA factories' productivity increased by 84%, whereas the improvements in office workers' productivity were insignificant. These factors, coupled with the advent of low-cost microprocessors, perhaps, are among the major reasons for the management to accept automation of office work.

The CRT display terminals have become almost as commonplace as typewriters in less than a decade. It is estimated that in Canada at least 100 000 workers spend most of their working day using CRT display terminals. A brief description of CRT display terminal-related tasks has been published (Grover 1976). Editing of text using a CRT display terminal is becoming more and more acceptable in every office.

The text to be edited is presented to the editor at a given rate known as baud rate (number of bits presented per second). The editor carries out necessary corrections. There are numerous variables that may effect the performance. These variables can be CRT display terminal-related, environment- and task-related, and above all worker-related. A significant review of most of these factors has been published (Cakir *et al.* 1980, Grover 1976, Machover 1968, O'Hare 1971). Factors influencing the CRT display terminal output are shown in figure 1.

Managers, task designers, union officials and others are experiencing a dearth of data and research findings in the literature pertaining to work

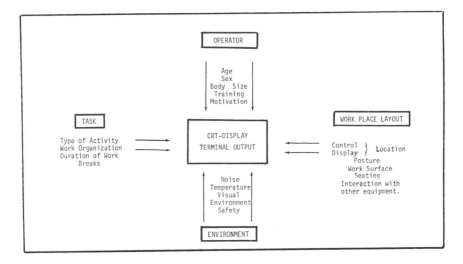

Figure 1. Factor influencing CRT display terminal output.

organization, rest pauses and output standards for such tasks. Some of the workers compare pacing of CRT display terminal-oriented tasks with assembly-line tasks. To increase our understanding of pacing on editing tasks, a set of experiments were undertaken.

2. Method and subjects

The text selected for editing consisted of 49 lines, each of which had an average of 43 characters and contained one error. Errors were underlined on the text, which was left in front of the subjects while editing. This text is shown in Appendix A.

A special package available on PDP 11/60 computer (Finar) was used as the word processor. This package was primarily chosen because it was available in the Computer Center and it had editing features which were relatively easy to understand.

Four terminals were used in the experiment. The particulars of the four terminals are as follows:

(1) Televideo 1–912;
(2) Digital VT55–EA;
(3) Lanparscope XT50;
(4) Volker-Craig VC–404.

These terminals apparently are very similar to each other, with the exception of VC–404. In this particular CRT display terminal, the keyboard is not an integral part of the unit. A table showing the important features of the terminals is shown in Appendix B.

Ten subjects participated in the study. These subjects had a limited experience in the use of CRT display terminals but had no significant experience in text editing. The subjects were interested in the experiment. They were student and office workers with an average age of 28 years. Apparently, none of the subjects suffered from any physical handicaps.

Out of several text editing features of Finar the following three were used in this experiment:

(1) delete character from a line!
(2) to insert character in a line ⟨
(3) to replace character from a line, type correct character at the required place.

The subject's task consisted of typing the text line number. This resulted in the appearance of the line on the screen. He was required to move the cursor to the error location and take the necessary action. The error was underlined on the text which was in front of the subject.

Time taken by the subject for completing the text editing was measured using a standard stop watch. Errors made by the subject were also recorded.

Before data collection was started each subject was familiarized with the location and operation of the control keys by giving him a practice run of ten minutes.

For selecting each experimental condition a $5 \times 4 \times 10$ factorial design was used. The baud rates used were 300, 600, 1200, 2400 and 4800 bits/s.

3. Results

Variations in number of errors and time taken at each baud rate for all the terminals listed are shown in tables 1 and 2.

Analysis of Variance (ANOVA) for time as well as error shows that baud rate is a significant factor. CRT display terminals are not significant.

4. Conclusion

Results obtained in this experiment show that baud rates are highly significant. The time needed for completion of a particular task decreases as the baud rate increases. This, however, does not follow a linear pattern. This is due to the fact that part of the task is human-paced. In fact, based on the results obtained in this experiment, the time decreases at a diminishing rate after a baud rate of 1200 bits per second and levels off very fast for higher baud rates.

Table 1. Variation in time. Table of terminal by bauds. Time was measured in minutes. Data pooled for all subjects for each condition.

Terminal						
Frequency Percent Row % Col %	300	600	1200	2400	4800	Total
Volker-Craig	134	127	125	119	106	612
	5·32	5·04	4·93	4·73	4·21	24·23
	21·97	20·79	20·37	19·52	17·36	
	24·66	24·41	24·82	24·29	24·84	
Lanparscope	137	129	126	126	120	638
	5·43	5·09	4·99	4·98	4·74	25·23
	21·53	20·19	19·76	19·73	18·79	
	25·17	24·67	25·08	25·57	25·74	
Decscope	135	130	128	127	120	642
	5·35	5·16	5·07	5·04	4·77	25·39
	21·07	20·32	19·98	19·85	18·78	
	24·78	25·00	25·52	25·89	25·89	
Televideo	138	135	123	119	119	635
	5·48	5·35	4·89	4·72	4·71	25·14
	21·80	21·28	19·44	18·78	18·71	
	25·39	25·92	24·58	24·25	25·54	
Total	545	522	502	492	465	2527
	21·59	20·64	19·88	19·47	18·42	100·00

An optimum baud rate was found by superimposing the plots of time and errors percentage versus baud rate. This rate falls in the range of 1100–1200, at which point the two curves of time and errors intersect. This is shown in figure 2.

Although CRT display terminals are not a significant factor, an analysis of total number of errors made by subjects on all terminals has shown that the least number of errors are expected when working with keyboards which have a tactile feedback system available on them, which assists the operator by providing him with an audible feedback. This is very useful, since the operator is aware that the character has been accepted and there is no need to depress the key further, which in turn results in less finger movement and hence is faster and improves operator performance.

The results of this preliminary study, which is limited in scope, indicates that an optimal baud rate, which yields in the minimum error rates and minimum

Table 2. Variation in number of errors. Table of terminal by bauds.

Terminal						
Percent Row % Col %	300	600	1200	2400	4800	Total
Volker-Craig	19 4·80 16·52 33·93	23 5·81 20·00 36·51	17 4·29 14·78 20·99	27 6·82 23·48 28·42	29 7·32 25·22 28·71	115 29·04
Lanparscope	15 3·79 15·46 26·79	18 4·55 18·56 28·57	20 5·05 20·62 24·69	21 5·30 21·65 22·11	23 5·81 23·71 22·77	97 24·49
Decscope	11 2·78 12·09 19·64	12 3·03 13·19 19·05	19 4·80 20·88 23·46	24 6·06 26·37 25·26	25 6·31 27·47 24·75	91 22·98
Televideo	11 2·78 11·83 19·64	10 2·53 10·75 15·87	25 6·31 26·88 30·86	23 5·81 24·73 24·21	24 6·06 25·81 23·76	93 23·48
Total	56 14·14	63 15·91	81 20·45	95 23·99	101 25·51	396 100·00

editing time, is nearly 1200 bits/s. Further analysis of the data collected, and extended experimental studies using more editing features are in progress. It is hoped that this study has increased our understanding of operator performance for such tasks. This understanding, of course, is a prerequisite for the development of adequate managerial control for such tasks.

Acknowledgements

The authors wish to express their appreciation of the financial support of the Office of Research Services, University of Windsor, for this investigation.

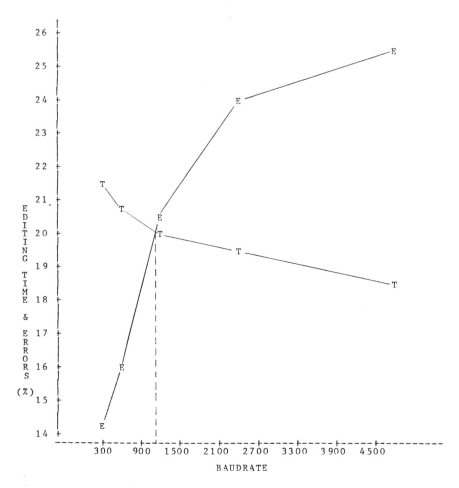

Figure 2. Time and errors versus baudrate.

References

CAKIR, A., HART, D. J., and STEWART, T. F. M., 1980, *Visual Display Terminals: A Manual Covering Ergonomics, Workplace Design, Health and Safety, Task Organization* (New York: John Wiley).

FINAR, *A Financial Analysis and Reporting System, User's Manual*: Finar Systems, 280 Riverside Drive #13H, New York, NY 10025.

GROVER, D., 1976, *Visual Display Units and Their Applications* (I.P.C. Business Press).

MACHOVER, C., July 1968, CRT Display Technology. Profiles, *Modern Data.*

O'HARE, R. A., July 1971, Interactive CRT Terminals, *Modern Data.*

Appendix A

Text used for editing

100. OPTIMIZATION MEANS FINDING THE BEST OF NANY
101. POSSIBLE SOLUTIONS. COMPUTEER OPTIMIZATION
102. TECHNIQUES BY WILLIAM CONLEY DEVELOPS A GREETLY
103. SIMPLIFIED APPROACH TOO THE USE OF COMPUTERS FOR
104. THIS PURPOSE. TH_COMPUTER TECHNIQUES EXPLAINED
105. AND ILLUATRATED HERE MAKE OPTIMIZATION AVAILABLE
106. TO PROFASSIONALS_HO HAVE LITTLE OR NO TIME TO
107. DEVELOP THEORETICAL EXPERTISE INN MATHEMATICAL
108. PROGRAMMING. IN FACT BASIC ARITHMETRIC IS ALL YOUE
109. NEED TI FULLY GRASP AND UNDERSTAND THE MATERIAL.
110. HANDS-ON BASIC WITH PET B?Y HERBERT D. PECKHAM
111. EXPLAINS HOW TOO 'PROGRAM FOR THE PET PERSONAL
112. COMPUTER USING BASIC. THI GUIDE FULLY EXPLAINS
113. THE APPROACHES ANDD TECHNIQUES NEEDED TO
114. SUCCESSFULLY USE THE NOST POPULAR COMMODORE PET
115. COMPUTER. SFTER A BRIEF STATEMENT OF OBJECTIVES
116. EACH CHATER PROVIDES A STEP BY STEP
117. PRESENTATION OFF KEY PROGRAMMING CONCEPTS.
118. DISCOVERY EXERCISES ILLUSTRATE THEPOTENTIAL OF
119. BOTH PET QND BASIC.
120. NI ONE DENIES THAT THE BALANCE OF MILITARY POWER
121. HAS CHANGED SINCE THE U.S. LAST DIRCTLY
122. CQNFRONTED THE SOVIET UNION WITH AA DISPLAY OF
123. FORCE IN THE 1962 CUBAN MISSILE CRISI. IN THE
124. PAST TWN YEARS ALONE, THE SOVIETS HAVE SPENT $150
125. BILLIONS MORA THAN THE U.S. ON THEIR MILITARY
126. FORCES. THWY HAVE ACHIEVED NUCLEAR PARITY WITH
127. THE U.S. QND MANY CONVENTIONAL WEAPONS. THE
128. SOVIETS OUTNUMBER THE U.S. 2 TO 1 INN MANPOWER.
129. DURING THE PAST YEAR, HOUSING HASS TAKEN SOME
130. OF THE HEAVIEST BLOWS FRIM THE
131. RESESSION. MORTGAGE RATES SOARED TO A HIGH OF 17%
132. IN SOMEE AREAS. AS A RESULT, HOUSING STARTS,
133. WHICH TWO YEARS AGO WERE RUNNING AT A_ANNUAL
134. RATE OF 2 MILLIONS, DROPPED TOO 900,000.
135. WHEN INTEREST RATES ON HOME LOANS FELL TO QBOUT
136. 12%, CUSTOMERS AGAIN BEGAN SHPPING FOR NEW
137. HOMES. THE SLUGGIST HOUSING MARKWT IS SURELY A
138. BAD OMEN FOR THE CURRENT HEALH OF ECONOMY.
139. THE CRT DISPLAY TERMINAL HAS BECOME ALMOST ASS
149. COMMONPLACE AS THE TELETYPEWRITER IN LESS THEN A
141. DECADE, FINDING ITS WAY INTO INUMERABLE
142. APPLICATIONS. *WHEN FIRST INTRODUCED, THE CRT
143. TERMINAL FOUND LIMITED ACCEPTANCE BECAUSE OFF
144. HIGH COST. PRICES HAVE STEADILY DECLNED, AND
145. SOME CRT'S ARE NOW LESS EXPENSIVE THEN THEIR
146. TELETYPEWRITER EQUIVALENTS. OVET 200,000 CRT
147. TERMINALS AR IN USE, COMPRISING ABOUT 30% OF
148. ALL DATE COMMUNICATIONS TERMINALS EMPLOYED TODAY.

Appendix B
Particulars of CRT display terminals tested

	Terminal and model			
	Televideo TV1-912	Lanparscope XT-50	Decscope VT55-EA	Volker-Craig VC-404
Size of keyboard (cm × cm)	37 × 9	38 × 10	35·9 × 9	32 × 9
Size of screen (cm) (diagonally measured)	29	30	30	28
Character generation method	← dot matrix →			
Cursor's shape	▢	▢	(blinking type) ▭	▢
Type of keys	← stepped →			
Audible feedback	No	No	Yes	No

VDT-computer automation of work practices as a stressor in information-processing jobs: some methodological considerations

By S. L. SAUTER, G. E. HARDING, M. S. GOTTLIEB and J. J. QUACKENBOSS

Department of Preventive Medicine, University of Wisconsin, USA

Application of video display terminals (VDTs) in information-processing work may introduce stressful job-design changes sometimes associated with the mechanization and machine regulation of work. This chapter identifies some of the more unique stressors linked to working with VDTs, and discusses an appropriate methodology for studying relationships among VDT work and workplace design features, and indices of worker wellbeing.

1. Background

Information-processing work is rapidly advancing as a major form of employment. Some estimates have it that by 1990 this type of work will engage up to 40% of the US workforce (Byron *et al.* 1980). Until the relatively recent introduction of 'intelligent' electronic instrumentation into office places in an effort to improve paperwork efficiency, information processing has not experienced the forces and effects of mechanization and automation which have impacted most other industrial processes. This technological revolution, represented mainly by the rapidly growing use of VDTs, now presents clerical workers with the first major job-design changes ever experienced in their field.

Concern has been expressed that the proliferation of computer automation in offices is introducing to office jobs many of the stressors associated with the mechanization and automation of routine industrial processes, or exarcerbating problems already experienced to a degree in clerical work (Nussbaum 1980). Many general parallels between the functional aspects of machine-controlled assembly-type industrial work and VDT-computer·automation of routine information-processing work support to a degree some of these projections. Perhaps the strongest link between these two types of work is that general latitude of control by the individual operator over many parameters of the job is diminished. In addition, work tasks tend to become more specialized, subdivided and regimented, and the implementation and opportunity for enforcement of more stringent performance standards is increased.

There exist, however, several interrelated conditions of VDT-computer automated information-processing work which distinguish it both qualitatively and quantitatively from most mechanized industrial work practices, and which may have the effect of further promoting job stress.

(1) Task performance is achieved through a reciprocal 'yoked' relationship with a 'smart' machine which may either present or control the task directly, as in updating or processing information from storage, or may function as a tool (input device) in accomplishing an external task such as coding paper records. The practical outcome of this arrangement is that VDT operators performing routine work often feel more tightly controlled by external machine systems than in conventional machine-regulated work where the relationship between the machine and the operator is more open-loop.

(2) Decision-making in task performance is more relegated to the machine (computer), and in job design to systems programmers who are more familiar with the constraints of the machine than of the operator.

(3) Temporal factors: slowdown caused by heavy use and machine malfunction interferes with the pace required to meet production standards. Allowance for a degree of self-pacing within the constraints of machine pacing is absent, i.e., real-time sensing and control of production does not enable the operator to 'work ahead' nor to accumulate work.

(4) Workers may have a very limited understanding of the equipment they work with or of the task itself, which may be highly symbolic or abstract in nature.

(5) Monitoring of operator performance in terms of output (e.g. key strokes) per unit time or other production indices is greatly facilitated.

Although considerable attention has been paid to the effects of physical stressors posed by the operation of VDT-computer automated information-processing work (Cakir *et al.* 1980, Grandjean and Vigliani 1980), there has been little systematic study of how job organizational and psychosocial changes caused by the introduction of this technology affects worker well-being. National Institute for Occupational Safety and Health (NIOSH) investigators (Smith *et al.* 1980) were the first in the USA to show an association in VDT work of psychosocial and job-task design factors with a broad spectrum of somatic and psychological complaints. Consistent with expectations based upon prior studies of job stress related to machine pacing and control of work (Caplan *et al.* 1975, Frankenhaeuser and Gardell 1976, Johansson *et al.* 1978), conditions suspected as prominent stressors were job-future uncertainty, loss of work identification and autonomy, workload and time pressure, as well as physical VDT-related and environmental factors. These results supported previous findings by Gunnarsson and Östberg (1977) who examined the complaints of VDT operators for the Scandinavian Airlines System and reported lack of job-content variety, time pressure, loss of participation and control in task planning or performance, and the need for constant vigilance, as major dissatisfactions. Cakir *et al.* (1980) identified feelings of loss of individual control and job alienation as a major consequence of VDT-computer automation of clerical-type work, and found these perceptions strongly related to reports of fatigue, monotony, stress, and loss of security and job meaning among VDT operators. These effects were amplified in VDT workers with the most routine jobs, and were most severe in clerical staff 'dequalified' to VDT positions.

Despite these data, the exact connection between VDT job-design factors and job stress is unclear. A 1978 New Zealand survey of 257 VDT operators and 124 control subjects found no meaningful group differences in estimates of pressure at work, satisfaction with work challenge or content, the degree to which the job was considered boring or frustrating, and headaches at work. A major problem in understanding the stress linked to machine control or pacing of work is discerning the separate and interactive effects of the many work and workplace design problems commonly found under these conditions. In studying social psychological and neuro-endocrine reactions in highly mechanized work, Johansson *et al.* (1978) identified 19 significant job content differences between the high-risk (mechanized-work) group and the control group. Although the investigators stated that it was not possible to draw valid conclusions about the effects of single job factors, univariate correlational techniques were used to infer the importance of specific conditions as important stressors. Likewise, machine-paced assembly-line work shown by Caplan *et al.* (1975) to be more stressful than non-machine-paced assembly work or other occupations differed from the comparison jobs in multiple job dimensions.

The fact is that machine pacing or regulation of work invokes a host of job conditions or demands which tend to co-vary and combine in complex ways to produce stress. This is nowhere more true than for VDT workers. In research we are conducting for NIOSH on stress related to VDT operation, we have observed a strong co-variation among deficiences in job design as well as physical environmental problems. Most notably, these include increased production demand, increased machine control of workpace, decreased decision-making in task performance, job-future uncertainty, decreased participation in job decisions affecting the worker, and lighting and work-station physical design faults. Without more specific information on the relative importance of these work conditions and their interactions in producing job dissatisfaction and stress, it is impossible to address the very important practical issues of how to design VDT work and workplaces to avoid the problems now being experienced by some operators, and the reason for conflicting data on the stressfulness of VDT work.

In our present research we are attempting to obtain this information through the use of the multivariate technique of multiple curvilinear regression. This procedure has been widely used in epidemiological research which is similar in methodology to our field study of job stress. The main advantage of this approach over univariate procedures is that it enables quantitative measurement of the strength of the relationship of any single job variable or combination (interaction) of variables with an index of stress or any other measure of interest while controlling for the influence of all other variables considered. This makes it possible to extract crucial variables and their interactions from extraneous or spurious ones in studying job stress in a way which is impossible with univariate methods.

The power of multiple regression as applied to the study of the relationship between job characteristics and job stress has recently been demonstrated by Karasek (1979). Using data obtained both in Sweden and the USA, he was able to show that job-decision latitude and other job demands interacted such that the negative consequences of demanding work can be offset if the worker maintains control or discretion in performance of the task.

2. Current research

In our research in progress, a questionnaire has been administered to 250 VDT operators and 84 clerical workers who manually performed tasks similar to those of the VDT group. Most participants are women, and public employees of the State of Wisconsin. No meaningful demographic differences exists between groups, although variation exists within the VDT group with respect to the nature of the VDT task performed and amount of time working at a VDT daily. The use of VDTs is not a major labour issue for these employees. Information obtained with the questionnaire includes the following general subjects:

(1) job descriptions;
(2) perceived job stressors;
(3) characteristics of the physical and social work environment;
(4) description of work tasks and equipment;
(5) health habits, status and history, including emotional indices;
(6) social and economic characteristics of participants.

Included are scales devised by Caplan *et al.* (1975) to evaluate job demands, by Insel and Moos (1974) to measure parameters of the physical and psychosocial work environment, and by McNair *et al.* (1971) to assess mood states.

For the ongoing analyses, measures comprising 15 constructs of major relevance to the wellbeing of workers and potentially influenced upon the introduction of VDT automation of information processing have been derived:

(1) *Control*—measure of the extent to which the worker experiences autonomy or control over job-pace or schedule, participation in job-task decision-making, task variability, and an absence of externally imposed performance standards;
(2) *Workload*—quantitative workload index;

(3) *Environmental*—measure of temperature, light, noise and other workplace physical problems;
(4) *Age* of the worker;
(5) *Eyewear*—whether or not worker wears corrective eyewear;
(6) *Self-perception*—measure of worker self-perception in terms of importance, success, and utilization of skills;
(7) *Bad habits*—measure of destructive personal habits and coping style;
(8) *Support*—measure of social support for the worker;
(9) *Job future*—measure of job-future certainty/security perception;
(10) *Job dissatisfaction*—general job dissatisfaction measure;
(11 *Job stress*—measure of perception of stressfulness of job;
(12) *General morbidity*—general morbidity index;
(13) *Skeletal muscular*—skeletal and muscular symptoms index;
(14) *Neuro-psychological*—neuro-psychological symptoms index;
(15) *Visual-ocular*—visual-ocular symptoms index.

Stepwise multiple curvilinear regression (Draper and Smith 1966) is being utilized to determine which of the variables 2–15 and/or their interactions are significantly related with the degree of control experienced by the participants in their jobs. The control factor was chosen as the dependent or 'criterion' variable for the analysis, since personal control in work is a construct which perhaps more than any other condition signifies or embodies the major difference between manual and VDT-computer automated clerical work. A 'dummy' variable consisting of VDT use versus no VDT use will then be interacted with each of the other variables in order to determine if it is justifiable to conduct further analyses for VDT users and non-users separately.

Acknowledgments

This report is based upon research carried out under contract with the US National Institute for Occupational Safety and Health, Motivation and Stress Section. Dr. Michael J. Smith and Dr. Alex Cohen played a major role in the design of this work.

References

BYRON, C., GELINE, R., and RAFFERTY, S., Nov. 7, 1980, Now the office of tomorrow. *Time,* 81–83.
CAKIR, A., HART, D. J., and STEWART, T. F. M., 1980, *Visual Display Terminals* (Chichester: John Wiley).
CAPLAN, R. D., COBB, S., FRENCH, J. R. P., Jr, VAN HARRISON, R., and PINNEAU, S. R., Jr, 1975, *Job Demands and Worker Health* DHEW (NIOSH) Publication No. 75-160 (Cincinnati: National Institute for Occupational Safety and Health).

360   *S. Sauter* et al.

DRAPER, N. R., and SMITH, H., 1966, *Applied Regression Analysis* (New York: John Wiley).

FRANKENHAEUSER, M., and GARDELL, B., 1976, Underload and overload in working life: Outline of a multidisciplinary approach. *Journal of Human Stress,* **2,** 35–46.

GUNNARSSON, E., and ÖSTBERG, O., 1977, *The physical and psychological working environment at a terminal-based computer storage and retrieval system, Report 35,* Stockholm, The Swedish National Board of Occupational Safety and Health, Department of Occupational Medicine.

GRANDJEAN, E., and VIGLIANI, E. (editors), 1980, *Ergonomic Aspects of Video Display Terminals* (London: Taylor & Francis).

INSEL, P., and MOOS, R., 1974, *Work Environment Scale-Form S* (Palo Alto: Consulting Psychologist Press, Inc.).

JOHANSSON, G., ARONSSON, G., and LINDSTRÖM, B. O., 1978, Social psychological and neuroendocrine stress reactions in highly mechanized work. *Ergonomics,* **21,** 583–599.

KARASEK, R. A., Jr, 1979, Job demands, job decision latitude, and mental strain: Implications for job redesign. *Administrative Science Quarterly,* **24,** 285–311.

MCNAIR, D., LORR, M., and DROPPLEMAN, L., 1971, *Profile of Mood States* (San Diego: Educational and Industrial Testing Service).

New Zealand Department of Health, 1980, *Video display units: A review of potential health problems associated with their use,* Wellington, New Zealand.

NUSSBAUM, K., 1980, *Race against time: automation of the office* (Cleveland, Ohio: Working Women).

SMITH, M. J., STAMMERJOHN, L. W., COHEN, B. G. F., and LALICH, N. R., 1980, Job stress in video display operation. In *Ergonomic Aspects of Video Display Terminals,* edited by E. Grandjean and E. Vigliani (London: Taylor & Francis).

Section 8. Problems in determining the relationship between production work and stress

Current status and research needs in occupational stress with special reference to machine-paced work[†]

By G. Salvendy[1], M. J. Smith[2], B. B. Morgan[3], T. K. Sen[4], T. J. Triggs[5] and M. Haider[6]

[1] Purdue University, USA
[2] National Institute of Occupational Safety and Health, USA
[3] Old Dominion University, USA
[4] American Telephone and Telegraph Company, USA
[5] Monash University, Australia
[6] University of Wien, Austria

Current concepts, measurements and modelling of human stress are reviewed in relation to machine-paced operations and computer-controlled work. Proposed priority research areas in the field of occupational stress are identified with added emphasis on machine-paced work.

1. Concept of stress

Depending on one's view of stress, the term can be regarded primarily as either an independent variable, an intervening variable, or a dependent variable. One proposed approach would be to call the stimulus variables stressors, the intervening variables stress and the response variables strain. Under this definition, one could examine stress in three basic ways.

2.1. *Quality versus quantity*
In normal life, most people encounter various kinds of stressors, some, by

[†] Following the International Conference on Machine Pacing and Occupational Stress, a one-day workshop was held on current status and research needs in the conference topics area. The workshop was divided for the morning session into the following four groups: (1) Measurement of Stress (Chairman: Ben B. Morgan); (2) Variables Relating to Stress (Chairman: Thomas J. Triggs); (3) Models of Human Stress (Chairman: Manfred Haider); (4) Impact of Machine Pacing on Stress (Chairman: Tapas Sen). In the afternoon, a plenary session was held in which the information derived from the four separate morning sessions was discussed. The material reported in this chapter is the product of the input of the workshop participants. The authors of this chapter acted only in an editorial capacity to integrate the discussions. The majority of those papers appear in this book have participated in the workshops. Each session chairman was requested to focus the group's discussion on the following: (1) summarize the current understanding in the field of discussion; and (2) given that research funding is available, identify the research which the group would like to do.

their very nature, being more harmful than others. Beyond a certain threshold, almost any kind of stressor could produce undesirable or harmful results (strain).

2.2. *Acute versus chronic*

Depending on the work condition and the worker, the stress manifestation could be a chronic or acute problem. Acute problems are limited to short-term exposures of a limited consequence, while the chronic problems are long-term daily exposures with cumulative effects.

2.3. *Short-term versus long-term*

Job demands may produce either a short-term or long-term stress condition on the individual. The short-term demands typically produce minimal effects, since the body has a chance to recover. The long-term effects tend to produce serious health problems such as coronary heart disease.

2. Measurement of stress

The literature often appears to be chaotic because past researchers have measured stress in terms of response variables that are of particular interest to themselves; there often appears to have been little concern for the application of findings to other studies. This makes it very difficult to interpret, compare, or apply these findings. Thus, one of the major needs related to the measurement of stress is the availability of a general framework around which results can be organized and interpreted. Indeed, it appears that the major weakness of research in this area is not related to the available measures of stress-related responses, but to the ability to interpret, generalize, and apply specific results in the broader context of other major findings in the area.

It is important to recognize that an individual's coping mechanism is defined by the pattern of the various types of response, and that under certain conditions the coping pattern of some individuals may be different from that of others. Thus, multiple types of response should be measured in future research, and the interrelationships that exist among the various types of measure should be documented.

The need to further examine the pattern of stress-related responses suggests at least three research strategies which should be considered by future researchers. First, it suggests the need to use a variety of different measurement methodologies. That is, it is not sufficient to design a research programme around the requirements of a given assessment methodology or a given technique for measuring psychological or physiological changes. Rather, future research should incorporate the requirements of research procedures from at least two, if not all four, of the classes of stress responses. These stress responses include the following indices: (1) physiological, (2) psychological,

(3) performance, and (4) safety and health.

Second, it suggests the need to engage in *interdisciplinary* research. A given researcher may not be experienced in applying all of the above techniques. Thus, future research programmes should combine the expertise of several researchers who are familiar with various measurement procedures. Such *interdisciplinary approaches to stress* research will provide for the fine-grained analysis of the total pattern of responses by individuals who are capable of interpreting their findings in terms of possible relationships to other stress responses.

Finally, the examination of patterns of responses will require the application of different statistical and analysis procedures than have been traditionally used. Methods such as regression analysis, factor analysis, and response-surface analysis have been applied with some success in other areas. The increased use of various time-series analyses is important in order to further examine the time-dependent relationships among variables.

One of the most pervasive findings from the measurement of stress is that there are likely to be wide individual differences in response to a given stress condition. There is some debate concerning the need for additional research in this area, and there is certainly little need to continue research just to document the occurrence of individual differences. However, it appears that additional research is needed in order to determine (1) the extent to which an individual's pattern of responses is the same under repeated exposure to the same stress (or to different stresses) and (2) the reasons why individual response patterns are different. The collection of such data will have important implications for job selection and for the design of jobs, working conditions, and stress-management paradigms.

Another major research need identified by this conference is for additional longitudinal research. Future researchers should engage in programmes to investigate the frequency, intensity, duration and changes in patterns of stress-related responses across time. Such research should include indices from each of the types of response identified above and should use multivariate procedures to examine the time-related changes in the relationships among these variables. The need for additional longitudinal research is urgent.

The participants at this conference saw relatively little need for research related specifically to the development of new measurement techniques. As indicated above, researchers in this area seem to be relatively satisfied with the available measures, techniques and procedures. It appears that some additional research is needed to (1) develop scales for the measurement of psychological responses, (2) increase the accuracy and standardization of biochemical assays, and (3) establish the sensitivity, validity, and generalizability of available measures. However, the major needs for future research in the measurement of stress call for the systematic application of available procedures to measure the patterns of *responses across time* and to establish the nature and the antecedent causes of different response patterns by different individuals at different times under different conditions. It was generally felt that the areas of stress research do not give sufficient attention to investigating a reasonable range of the independent variables involved. There is a need to ensure that the region of possible interest is tapped and that functional relationships are validly defined (such as 'U-shaped' functions). There are a number of classes of independent variables that can be used to characterize performance under stress.

Task variables. This category would include such factors as the type of pacing, form of human-operator response, level of cognitive demands and decision-making.

Individual difference variables. Humans differ anatomically, physiologically, intellectually and emotionally. The tendency of an operator to develop stress responses in particular selection of individuals can improve systems performance. Alternatively, it is also desirable that systems be designed so as to accommodate a reasonably wide range of individual differences.

Environmental variables. Several dimensions are important, for example physical, social and management environments.

State variables. The state of the operator will be influenced by such variables as the amount of sleep loss, the time on task, and duration of the stress.

Training variables. The form and level of training is likely to affect the coping strategies. Special training might be able to alleviate the effects of stress.

Procedural variables. This class of variables relates to how the task components are organized. Restructuring of the order in which subtasks can be performed may provide a means of alleviating stress.

These classes of variables highlight the need to be concerned with the interaction effects between variables, particularly variables from different classes. In this overall area, where there is an imbalance between performance capacities and the situational demands, *interactions* are likely to be of more practical and theoretical significance than main effects.

While there was little interest in the development of new stress-measurement techniques, there was abundant interest in trying to define 'stress'.

3. Models of human stress

It has been suggested that the time stress only should be used as a label of a broad area of interest, and at the same time specific models should be built with more fully specialized terminology.

More specifically, an interesting model proposed was by Andries Sanders. The notion of certain variables influencing different stages or levels has appeal. The model appears to be sufficiently well formulated that systematic selection of variables from the task variable, state variable, and individual difference categories could be carried out so as to identify different patterns of performance. Analysis of such patterns could assist in validating the structure of such a model. Among these are useful models to explain the costs to the individual in maintaining sustained levels of performance in situations of boredom and understimulation as well as overload. To this extent, the concept of effort as defined by various authors proved to be very helpful. In model-building we have to accept that there is no simple dimension of 'stress' or of 'activation', but rather a complex of variables which may result in conflicting response patterns such as declining control arousal while increasing muscle tension. Such conflicting response patterns may have negative consequences for wellbeing and health which some people call 'occupational stress'. On the other hand, some of these complex interacting variables may be taken as challenges or demands for action which then, in turn, lead to positive consequences, for example learning experiences under circumstances of challenge and collective social actions to change adverse job situations.

Issues have been raised during the conference relating to job content and task structures as factors interacting with more traditional physical aspects studied by economic analysis. In the case of video display terminals, the method of computer-system operation may significantly affect job content and strain response. This raises the possibility of 'Ergonomics of Software' as an area of future research.

A *taxonomy* of task variables appropriate to the area of stress is generally lacking. Such a taxonomy needs to be based on both theoretical considerations arising from models of stress and the requirement to investigate specific task environments.

4. Human aspects of machine-paced operation and computer-controlled work

Machine-paced work, like stress, needs some definition, given the variety of paced-work systems and meanings ascribed to 'paced'. Temporal control is considered to be the primary determinant of pacing, the more specific measures being cycle time and duration of work spell. Other secondary determinants of pacing are:

(1) Type of job;
(2) Type of capacity required, including training and demographic considerations;
(3) Task criticality;

(4) Autonomy or control by workers, both external and internal; and
(5) Individual versus group task condition.

All these, depending on their nature and extent, may cause occupational stress.

Basically, there are three broad areas where stress affects workers: (1) Physical health; (2) Emotional/psychosocial wellbeing; and (3) Productivity/performance. Some specific measures in each of these areas are as follows.

Physical health.
Mortality   Blood pressure
Morbidity   Heart rate
Pathophysiological reaction   Neurological
Muscle tension   Immunosuppressor reactions
Hormone reactions

Emotional/psychosocial wellbeing.
Ancillary behaviour   Psychoneurotic reactions
Mood state   Psychoses

Productivity.
Quantity produced   Accidents
Quality of product   Arbitrary work breaks
Absenteeism   Physiological and psychological cost
Turnover

Unlike the physical-health indicators, productivity cannot be used as a direct measure of stress. However, the cost of productivity or performance degradation as a result of stress is certainly an important subject that needs further investigation.

It is anticipated that future trends in new technology will be towards more computer-controlled work environments. These will generate more mental demands for workers, leading to various kinds of occupational-stress phenomena that differ from those on the shop floor. This anticipated stress will have a high probability of undesirable impact on physical health, emotional/psychosocial wellbeing as well as on the productivity and performance of the workers. It is extremely important that we develop a better understanding of what kinds of stress this machine-paced environment might produce at the workplace and what the potential social costs would be. With this knowledge on hand, one would then be able to design future technological systems to be more efficient and to have minimal impact on worker health, work quality, and productivity.

On the assumption that the trend in new technology is towards more computer-controlled systems which will in turn demand more mental and monitoring types of task, the following three major research activities were identified as crucially important for the future:

(1) Define the needs and capacities/limitations of the human in the new system;

(2) Articulate and enumerate what we already know about stress and its relationship to work variables in today's technological system; and

(3) Define the gap that may exist between existing knowledge and the data/information we need for more effective design of new technology.

Index